Inspecting and Diagnosing Disrepair

Inspecting and Diagnosing Disrepair provides housing officers, surveyors, landlords, tenants, lawyers and environmental health inspectors with the essential information they need to record, diagnose and remedy disrepair. Pat Reddin presents technical information methodically, including useful diagrams to help readers to develop an understanding of building materials and structures and to advise and take action on disrepair. The book is fully up to date with the latest legislation and is essential reading for environmental health professionals, surveyors and students alike.

Patrick Reddin (1947–2015) was a Director of Reddin & Company Limited, Chartered Building Surveyors and Corporate Building Engineers, London, and Executive Director of ABE Enterprises. He was also a Past President of the ABE.

Inspecting and Diagnosing Disrepair

Patrick Reddin

LONDON AND NEW YORK

First edition published 1996
by Lemos & Crane

This edition published 2016
by Routledge
2 Park Square, Milton Park, Abingdon, Oxon OX14 4RN

and by Routledge
711 Third Avenue, New York, NY 10017

Routledge is an imprint of the Taylor & Francis Group, an informa business

© 2016 1996 Patrick Reddin

The right of Patrick Reddin to be identified as author of this work has been asserted in accordance with sections 77 and 78 of the Copyright, Designs and Patents Act 1988.

All rights reserved. No part of this book may be reprinted or reproduced or utilised in any form or by any electronic, mechanical, or other means, now known or hereafter invented, including photocopying and recording, or in any information storage or retrieval system, without permission in writing from the publishers.

Trademark notice: Product or corporate names may be trademarks or registered trademarks, and are used only for identification and explanation without intent to infringe.

British Library Cataloguing-in-Publication Data
A catalogue record for this book is available from the British Library

Library of Congress Cataloging in Publication Data
Reddin, Patrick.
  Inspecting and diagnosing disrepair / Patrick Reddin.
    pages cm
  Includes bibliographical references and index.
  1. Building inspection. I. Title.
  TH439.R43 2016
  690'.24—dc23   2015016096

ISBN: 978-1-138-80208-7 (pbk)
ISBN: 978-1-315-75436-9 (ebk)

Typeset in Sabon
by Keystroke, Station Road, Codsall, Wolverhampton

This book is dedicated to Vincina Mellor and Andrew Arden QC

Contents

Foreword by Andrew Arden QC   xiii

Introduction   1

PART I
Understanding buildings   5

1   Housing stock   7
TRADITIONAL   7
External walls   8
   Solid brickwork   8
   Cavity brickwork   8
   Blockwork   9
Internal walls   9
   Stud partitions   9
Foundations   9
   Timber plates   9
   Brick footings   10
   Strip foundations   10
   Raft and other reinforced concrete foundations   10
   Piling   10
Roofs   10
   Pitched roofs   10
     Valley roofs   11
     Front to rear pitched roofs   11
     Hips, valleys and gables   11
     Parapets, fire walls and chimneys   11
     Slates and tiles   11
   Flat roofs   12

Floors   13
   Above ground level   13
   Timber ground floors   13
   Solid ground floors   13
   Boards and sheets   13
Damp-proofing   14
   Walls   14
   Ground floors   14
   Basements   14
Windows   14
   Sliding sashes   14
   Casements   15
   Replacements   15
NONTRADITIONAL   15
Steel and reinforced concrete frame   15
Large panel systems   16
No-fines concrete   16
Modern timber frame   16

2   Materials of construction   18

Timber   18
Brick   20
Mortar and pointing   20
Render   21
Insulation   21
Rock   21
Concrete   22
Plaster   22
Stone   22
Slate   23
Tiles   23
Glass   23
Metal   24
Asphalt, felt and other bituminous materials   24
Medium Density Fibreboard   25

3   The enemies of healthy buildings   26

Water   26
   Dampness from the ground   27
      Rising damp   27
      Lateral penetration   28

Remedial damp-proofing 28
    Bridging 29
    Salts and residual dampness 29
Above ground 30
From above 31
    Cavity construction for walls 31
From inside 31
    Condensation 31
    Interstitial condensation 32
From construction 32
Sulphates 32
Fungi 32
   Dry rot 33
   Wet rot 33
   Moulds 34
Insects 34
   Wood boring 34
   Disease carrying 35
   Just unpleasant 35
Metals 35
Hazardous materials 36
   Asbestos 36
      Removal contractors 37
   Glass fibre 37
   Urea-formaldehyde foam 37
   Radon gas 37
Users 38
   Moisture generation 38
Animals 38
   Disease carrying 38
Refuse 38
Plants, trees and bushes 39
Temperature and climate 39
   Drought 39
   Frost 39
   Snow 40
   Sun 40
Underground threats 40
Nontraditional buildings 40
   Prefabricated buildings 41
   Large panel systems 41

High-rises 41
Modern timber-framed buildings 42

PART 2
Inspecting disrepair   43

4   Preparing for the inspection   45
The housing manager's role   45
'Following the trail'   45
Starting points   47
  Occupied dwellings   47
  Voids   47
  Third parties   47
Is inspection necessary?   47
  Recording information   48
    Recording the initial information   48
Background information   49
  Address and identification code   49
  Type of property   49
  Size and location   49
  Building type   49
  Occupants   50
  Tenancy agreement   50
  Common parts   50
Equipment   51
  Basics   51
  Moisture meter   52
  Camera   52
  Use of drones   52

5   The inspection   53
Appointments and casual call systems   53
  Diary systems   54
Health and safety   55
Making appointments   55
Conducting the inspection   56
  Introductions   56
  Preliminaries   56
Methodical procedure   57
  Using the senses   58
  Identifying defects   58

Inspection notes 58
Timescales 67
Do-it-yourself 67

PART 3
Post-inspection practice   69

6 Reporting   71
Customers 71
The findings 71
Format 72
   Core section 73
   Supplementary sections 76
   Record note 76
   Landlord's liability 77
   Tenant's liability 77
   Reference to further action 78

7 Priorities   79
External advice 79
Specification for repairs 80
Assessing priorities 80
   General factors 80
   Statutory obligations 81
   Codes of practice 83
   Tenants' Guarantee 83
   Tenants' Right to Repair 84
   Chartered Institute of Housing standards 84
   Planned maintenance 84
   Wholesale redevelopment or rehabilitation 86
   Value for money 88

8 Follow-up action   91
Monitoring performance 91
   Management systems 91
Tenant satisfaction surveys 92
Redecoration 93
Variations of the works 93
Longer term action 94
   Monitoring short-term repairs 95
   Visits and revisits 95

xii Contents

9 **Court proceedings** 96
   Instructing an expert 96
   Expert evidence and advice 97
   Choice of expert 99
   Housing manager's evidence 100
   Compliance with orders 101

10 **Conclusion** 102
   Procedures 102
   Efficient use of resources 102

   Appendix I Building diagrams 105

   Appendix II Diagnosing building defects 121

   Index 124

Foreword

Remembrance of Patrick Bernard Reddin, 1947–2015

By Andrew Arden QC

Pat Reddin died after a fall on 13 April 2015, the day before his 68th birthday.

Pat did his training with the Inland Revenue Valuation Office and subsequently worked as a planning officer in local government before completing his finals in 1971 and co-founding Reddin & Nuttall in 1972; the firm later became Reddin & Co Ltd, and it would be the principal base for his career for the rest of his life. He was a Fellow of the Royal Institute of Chartered Surveyors, a Fellow of the Chartered Association of Building Engineers (CABE), a member of the Chartered Institute of Housing and of the Society of Expert Witnesses, and a mediator accredited by the Centre for Effective Dispute Resolution and by the Eastern Caribbean Supreme Court. He was particularly active in the CABE (previously the ABE) – he was president in 2001 to 2002 and honorary secretary from 2002 to 2009; he received its distinguished service award in 2010.

Pat worked and lectured in Latin America, the Caribbean and the Far East as well as in the United Kingdom. He was a visiting lecturer at Kings College London for its Masters in Environmental Health; and he conducted seminars for numerous bodies including the professional institutions of which he was a member, housing organisations, housing law associations and the Judicial Studies Board.

As well as authoring a number of articles for technical and professional journals and his membership of the Editorial Board of the *Landlord and Tenant Review* (Sweet & Maxwell), Pat's publications, as author or editor, include *Specifying Minor Building Works* (Routledge, 2013). In addition, he wrote – for a series I edited – *Dealing with Disrepair* (Arden's Housing Library, Lemos & Crane, 1996) of which this, his last book, *Inspecting and Diagnosing Disrepair*, is the revised edition. As the first had been, the book's focus is housing; it is written in a way that makes it not only accessible and useful to the housing managers to whom it is directed but also an informed and indispensable guide to this subject for other building professionals, including surveyors and environmental health officers as well as housing advisers, whether legal or otherwise.

Receiving the typescript from his publishers shortly after Pat's death, I was surprised, honoured and deeply moved to find that he had dedicated the book not only to his wife but also to me.

Pat's work, either here or abroad, was not confined to housing; he had a substantial party wall practice as well as being involved in a wide range of activities for commercial and other organisations. His principal activity, however, was housing; and from the beginning of his practice in his own firm, he was the go-to surveyor of choice in housing disrepair cases – both for tenants and for social landlords. He was widely used as a joint expert, reflecting both his exemplary standing and reputation for integrity within the legal profession and his own passionate commitment to housing. Pat's first allegiance was to the have-nots, the least well resourced, those most in need of help to achieve a decent home and a decent chance; it was an allegiance served not with fierce advocacy alone but with unrelenting, rigorous application of his skill to the task at hand.

This description does not adequately reflect Pat's huge professional contribution to the development of housing rights, starting in the 1970s and continuing. Lawyers and other advisers could and did develop a theory of housing law and a body of housing rights, whether through litigation or legislation; but these do not come to life without form, and where they concerned housing standards, it was the work Pat did – more than any other surveyor – that achieved this. When we went to court, it was not the legal argument that mattered nearly so much as the body of his report; no matter how easily – or even with what difficulty – a judge could brush aside our arguments, they simply could not ignore what Pat had to say.

I cannot write wholly objectively about Pat: he was my close friend for 40 years; we supported each other through our highs and lows, though I have no doubt that it is I who received the most support. Pat was an extremely generous, caring and giving man as witnessed in a hall packed beyond capacity at his funeral. Pat leaves behind not only a wife, four daughters and five grandchildren but also a wide range of friends and colleagues, the outpouring of whose love on that day is something that defies words.

In 2011, Pat was diagnosed with a brain tumour which he fought courageously, vigorously and (for the most part!) with the best of humour throughout 2012, allowing him to return to work less than a year after surgery. This was long before the doctors had predicted he would be able to do so, although there had been some real doubt for a while about both his chances of recovery and how far he was (or would be) capable of comprehension, particularly while he was still in the ICU after an 11½-hour operation in early January.

In my own mind, though, I was quite clear – even then – that I could see in his eyes flickers of recognition at some of the things that were said across his bed at the time; and it is one of the enduring joys of my life that as soon as he could communicate again, he confirmed that I had read him right: he had not left us but briefly.

That last thought struck home the day he died when I found a draft will he had written way back when – from its contents, probably 20 years or so before – and which contained this thought:

> This is a time of celebration and adventure. I am going on another journey. I will return and I will always be available for you. Hold me in your thoughts for that is where I reside.

To my dear gentle soul – I am with you always.
My wife

Introduction

This book is primarily for managers of rented housing. It helps them to identify and diagnose disrepair and to find remedies to deal with it. For tenants, the landlord's housing managers are the first point of contact. Increasingly, managers have to face complaints of disrepair and this book equips them to fully assess and deal with such problems.

Chapters 1 to 3 describe the various ways in which houses have been constructed, the qualities of materials used in buildings and the common causes of disrepair. Chapters 4 and 5 give the reader the tools and techniques for inspecting and diagnosing the common causes of disrepair. Chapters 6 and 7 explain how to report back on what has been diagnosed, where more professional help may be needed, and the legal obligations and financial constraints that often conflict with carrying out repairs work. Chapter 8 shows how to follow through and ensure that appropriate repairs have actually been done. Chapter 9 comes in where all else fails – covering the process of litigation in the courts. The concluding chapter emphasises methodical procedures and efficient use of resources.

There are increasing pressures on social landlords who are trying to sustain a well-maintained housing stock. Tenants' aspirations are rising. Government funding is falling. Much of the stock was badly designed to begin with and has already passed its planned lifespan. The demands are overwhelming, but failure to carry out repairs is even more costly – eating away at the capital value of the stock while generating expensive and time-consuming litigation on the way.

In response to these enormous pressures, the social landlord can at least ensure that disrepair is dealt with through top-quality management procedures. The modern business concept of quality has refocused relationships with external customers such as tenants. But it has also created new demands for professionals from different disciplines to work together effectively as a team within established written procedures, which are monitored in practice. All this is a prerequisite for those who strive for the European Standard of Quality Management Systems (ISO 9001), conceived as a means of helping to build the effective systems that this new outlook demands.

This book was originally published as *Dealing with Disrepair* by Lemos & Crane for Arden's Housing Library. *Inspecting and Diagnosing Disrepair* is essentially about structural disrepair and will look at and consider disrepair as it affects the occupation of the dwelling. It is not concerned with defects to services – i.e. heating systems, gas supply and electricity – except insofar as they have implications for the fitness and structure of the dwelling.

Structural disrepair has more than one meaning, however, because the word 'structure' is open to different interpretations. For architects, surveyors and engineers, something is 'structural' if it supports, strengthens or restrains the basic frame of the building – like load-bearing walls, floor joists and roof timbers.

In relation to disrepair, there is another way of looking at the meaning of 'structure'. In this view, every part of the structure – wall, roof, floor, and so on – is made of elements that only together make up a whole. Many walls have brick as their core, and this may be covered by a vapour barrier or insulation that in turn is covered by plaster, lining paper and paint. At some point, moving out from the structural brickwork, components are no longer structural but are simply part of the interior. This borderline is the most common demarcation between the respective repairing liabilities of landlord and tenant. This book explains how these elements fit together to form the whole of the structure.

There have been significant changes since this work was first published in 1996. The main changes are touched on in the text, but among the most substantial has been the move towards 'sustainable' construction, recognising the importance of energy efficiency and thermal performance in buildings. This has been largely achieved by changes to building control legislation which of course are not retrospective but, when taken with changes to the assessment of deficiencies in housing (the Housing Health and Safety Rating System, HHSRS), bring a new armoury to local authorities to achieve better standards. From 2006, Building Regulations have a requirement to carry out consequential thermal improvements to existing parts of a property, but this only applies when an extension is proposed and the existing property has a useful floor area exceeding 1,000m^2 (a large house). This is unlikely to change in the future.

However, the need to address energy use and such issues as cold and damp remains significant since such a large portion of the UK's housing stock is 100-plus years old and incapable of being upgraded at economic cost. The stage is therefore set for attempts to bring condensation-caused defects into the repairing requirements of landlords. However, it is difficult to see how this could be achieved without substantial government grant aid. Until the government assesses the true cost to the country of inadequate housing, it is not possible to assess the possible sums that could be made available for improvements from budgets currently allocated to health services and other areas that are impacted by housing defects.

> **Energy efficiency standards**
>
> There are proposals to raise energy efficiency standards in the private rented sector through staged legislation. The current draft legislation is The Energy Efficiency (Private Rented Property) (England and Wales) Regulations 2015.[1] The aim is to ensure that property (apart from certain exemptions outlined in the legislation) conforms to current minimum energy performance standards. There is also provision for tenants to make improvements: a tenant in a domestic property will be able to request the landlord's consent to make prescribed improvements; and consent should be given unless exemptions apply or the landlord sets out other measures. The tenant can appeal through the First-tier Tribunal.
>
> The legislation will be enforced by local authorities who will be able to issue compliance notices and impose fines on landlords failing to meet standards. Again, landlords will be able to appeal via the First-tier Tribunal.

In addition to developments in terms of the Building Regulations, the perspective of tenants has changed over time. Since the publication of the original book, many social housing tenants have seen standards raised through implementation of the Decent Homes Programme. This was introduced in 2000 with targets to improve standards by 2010, especially for local authority areas that experience higher levels of deprivation. In 2006 the standard was updated to take account of the Housing Act 2004 and the associated implementation of the HHSRS. The current criteria for the standard of social housing are as follows:

- be free of health and safety hazards
- be in a reasonable state of repair
- have reasonably modern kitchens, bathrooms and boilers
- be reasonably insulated.[2]

Unfortunately, the targets for the programme were not met and the Homes and Communities Agency now administers the Decent Homes Backlog Programme on behalf of the Department for Communities and Local Government. Nevertheless, the expectations of occupants in social housing have been raised.

Notes

1 www.legislation.gov.uk/uksi/2015/962/note/made?view=plain

2 https://www.gov.uk/government/publications/2010-to-2015-government-policy-rented-housing-sector/2010-to-2015-government-policy-rented-housing-sector#appendix-5-decent-homes-refurbishing-social-housing

Part 1

Understanding buildings

Chapter 1

Housing stock

Buildings are integral to the work of all those involved in the management of housing. The job titles and descriptions are myriad, and include housing managers, housing officers, maintenance staff and frontline housing staff. In this book, I use the term 'housing manager' as shorthand for all these roles.

Buildings are the main piece of equipment used by the manager to house the tenants. In order to appreciate and work with and even to enjoy the study of and the use of buildings, the housing manager needs to have an understanding of the various types of housing in the UK and the methods and components of their construction.

The UK housing stock is commonly divided into two categories: traditional and nontraditional. Until the late 1960s, this was an easy distinction to draw. The two categories were largely identifiable by their appearance without the need for analysis of background information. Repair problems and solutions were generally considered to be common to all types of buildings within each category. This was, regrettably, an oversimplification. Housing managers must go into the subject in more depth if they are to identify areas of risk for disrepair and come up with repair solutions.

This chapter describes the characteristics of both types of housing and the signs to look for to enable further identification within each group. The defects inherent in or likely to occur in each type are discussed in Chapter 2.

TRADITIONAL

We describe a building that is over 250 years old not as 'traditional' but as 'historic' or merely 'old'. Traditional housing is associated with the period following the great urban building boom that happened in the early and mid-Victorian eras, in the wake of the Industrial Revolution, when the once rural population came flooding into the towns. The first response to this demand from city dwellers was from the speculative private sector in the form of multi-occupied tenement blocks and rows of back-to-back terraced units in the very heart of our major conurbations.

The poor living conditions within these dwellings, with shared sanitary facilities and inadequate supplies of drinking water, soon gave rise to severe health problems and fears of a new plague.

Pressure from churches, charities and benevolent industrial owners resulted in the introduction of public health laws and the establishment of the social housing movement. This led to the construction of a vast number of dwellings in the latter part of the nineteenth century and the first 40 years of the twentieth century. These are now referred to as 'traditional' and this style continues to be built, although in declining numbers, to the present day.

The pressure to deliver dwellings speedily after the Armistice in 1918 fell to the public sector. A target of 500,000 dwellings in three years was set in the Housing and Town Planning Act 1919. Subsequent legislation in 1923 and 1924 produced another 503,000 units. The emphasis in the 1930s shifted to slum clearance; between 1935 and 1938, for example, the number of dwellings produced was 400,000.

However, if we examine the particular styles and methods of construction used over this period of perhaps 70 years, we find varied and significantly different qualities. These can be compared by looking at each element in turn and assessing the differences, which have significance in terms of current maintenance and repair. Any dates and periods given can only be rough guides. Innovation and varying trends throughout the UK resulted in widespread divergence of house styles and methods of construction.

External walls[1]

Solid brickwork[2]

Prior to 1900, walls were almost exclusively of solid brickwork. They were constructed of bricks laid one on top of another in a regular pattern to ensure that no joints ran vertically through more than one layer (course). The bricks were bedded in mortar, which incorporated lime to allow some flexibility for structural movement. With the mortar joint, the brick dimensions were 112mm (4.5") thick × 75mm (3") high × 225mm (9") wide. Solid brickwork was commonly 225mm (9") thick for buildings of two storeys; and where three or four storeys were constructed, this increased to 337mm (13.5") thick in the lower floors. Higher quality construction, for the merchant classes, commonly had a basement with 450mm (18") thick walls.

Cavity brickwork[3]

From about 1900, there has been increasing use of cavity wall construction and of masonry other than bricks. The use of cavity construction in the early part of the twentieth century is commonly found in Hampshire whereas elsewhere, for example, in London, it is rarely found in dwellings built before 1925.

Cavity walls comprise two skins of masonry, tied together with metal at regular intervals. The cavity is open at the top and bottom in order to allow air to circulate. It is assumed that in exposed areas, rainwater could breach the outer skin of brickwork. The ventilation of the cavity and the formation of twists and drips on the metal ties prevent transmission of moisture to the interior skin and keep the dwelling dry.

Blockwork[4]

The pressure to deliver large quantities of dwellings in the period from 1919 to 1938, together with a reduced and ill-prepared construction materials industry, triggered the more innovative manufacturers of materials and house designers to look at new alternatives to brick. The main innovation was the block. This is usually of the same thickness as the brick but measures 450mm (18") in length by 225mm (9") in height. Its larger size allows speedier erection, thus reducing labour costs and time.

Blocks became increasingly used for the inner skin of cavity walls. They were also used for exteriors that were to be rendered.

Internal walls[5]

Stud partitions[6]

Prior to 1925, the most common form of construction for internal walls was a timber stud partition. The timbers (studs) were arranged vertically at approximately 400mm to 500mm (16" to 18") centres with horizontal timbers (plates) at the top and bottom. Between the studs, timber noggins were installed to prevent lateral movement; and in more substantial construction, diagonal bracing was also included.

It is a common fallacy that these timber walls do not carry any loads. Invariably they are load-bearing, and the only safe assumption is that they are an essential part of the structure of the building until they can be shown to have no structural significance. From about 1925, internal walls, particularly load-bearing walls, were constructed of brick or block.[7]

Foundations[8]

Timber plates[9]

Early dwellings (i.e. pre-1800) were often built directly off the ground with a base layer of stones on which was laid a timber plate. Second-hand ship timbers were commonly used, the pitch applied for sea use giving good resistance to dampness and fungal or insect attack.

Brick footings[10]

Today, as a minimum, most rented housing stock will have brick footings. They comprise two or more courses of bricks, laid wider than the width of the wall on to broken stone or compressed ground to spread the load of the wall. The more courses and the greater the spread, the lower the pressure on the ground and therefore the less likelihood of failure.

Unfortunately, these footings are commonly shallow and, although adequate for distributing the load from the wall to the soil, are incapable of resisting the pressures exerted by tree roots, subsoil movement and changing water content.

Strip foundations[11]

To overcome the problems of ground movement, concrete strip foundations came into common use by about 1930. The concrete, usually about 300mm thick, is laid at a depth of about 1,000mm and is three times the width of the wall. The depth is to overcome the effects of frost on the ground. In many cases, these foundations are adequate but, as with brick footings, subsoil movement (such as settlement or subsidence) and tree roots can undermine and destabilise them.

Raft and other reinforced concrete foundations[12]

Where ground stability is known to be a problem, steel is introduced into the foundations to increase strength and elasticity. In some cases, rather than excavating to a great depth to find good ground from which to build, reinforced concrete slabs are laid, from which the superstructure can be erected. In effect, these rafts are deep, reinforced concrete floor slabs that spread the load of the superstructure over the site.

Piling[13]

Where ground conditions are unstable, piled foundations are used. These transfer the loads of the building deep into the ground on to sound bearings. The caps to the piles, which are reinforced with steel, are linked together either in a series of beams or as a slab (rather like a raft) off which the superstructure is built.

Roofs

Pitched roofs[14]

Pitched roofs are all those with sloping faces, draining down to a gutter. Almost without exception, the structures of these roofs are constructed of timber.

Valley roofs[15]

The Building Act of 1707 banned the construction of projecting timber eaves to roofs (because of the fire risk). As a result, in central urban areas – particularly in London – there was an architectural demand for front walls to be extended upward to form parapets. To employ a roof sloping down to the front of the house with an external gutter would have made this stylistic demand unattainable. The solution was the centre valley or butterfly roof. This comprises a central timber beam, often made up of two parallel timbers, supported on the front and rear external walls with part of the load transmitted down on to the central stud partition.

Front to rear pitched roofs[16]

This is the most common type of roof found in UK housing prior to 1945. The roof structure comprises rafters that are much like floor joists but which are laid to a slope. These are often supported by purlins – larger timbers running across the slope – that are, in turn, supported on struts down to the internal walls.

Hips, valleys and gables[17]

Where parts of the building are constructed on differing planes, the roof junctions are formed with hips (outer angles) and valleys (inner angles). Where roofs come to a vertical face, the triangular vertical panel is called a gable, regardless of its material.

All changes in direction and junctions are areas of increased vulnerability.

Parapets, fire walls and chimneys[18]

In modern construction, the front to rear roof on a terrace of houses is constructed separately for each house. Between the houses, the separating wall – usually of brick – rises above roof level as a parapet. In Inner London, building by-laws have required this parapet for almost 150 years; but outside this area, the wall commonly stops at the underside of the roof.

In terraced housing built before 1925, it was also common to leave out bricks in the separating wall within the loft. This permitted cross-ventilation throughout the whole terrace and was intended to minimise condensation and to remove minor dampness resulting from intermittent water penetration.

Slates and tiles[19]

Following the ban on thatch after the 1666 Great Fire of London, roofs in urban areas were covered with slate. Slate quarries in Wales and other parts of the UK were at their zenith in the mid-nineteenth century, with demand exceeding supply year-on-year. The inability of the slate quarries to satisfy demand led to the more widespread use of clay tiles, hitherto confined to localised manufacture. Large-scale tile production, alongside that of bricks and blocks, supplied a constant and adequate supply of material for the interwar building boom.

Tiles are not as durable as good-quality slate, but they are not dissimilar to the poorer slates produced at low price to satisfy the periods of heavy demand. Tiles were, and still are, cheaper than slate; and their supply seems inexhaustible since they are produced independently of the natural occurrence of suitable rock.

In the context of building defects, the primary distinction to be noted is the weight of the material and the consequent additional strength of the structure required to support it.

In the early 1970s, it was a common practice to renew slate roof coverings with concrete tiles. The increased weight necessitated additional support to the roof structure but this was not always provided. After a series of roof collapses, regulations were clarified, making the change in material a notifiable operation under the Building Regulations.

With the development of bituminous felts and plastic materials, particularly after 1945, under-slate/tile coverings began to be installed (these are a requirement in Scotland). These provide a secondary barrier to windblown rain and snow.

Flat roofs[20]

Flat roofs were rare before 1930 and did not come into common use until after the First World War. More recent additions to existing buildings may have flat roofs.

Flat roofs are less capable of shedding water than sloping roofs and are more susceptible to thermal movements. The coverings are therefore put under greater stress, causing earlier failure and reducing the life of the roof.

The coverings are of metal (usually zinc or lead)[21] or of asphalt or bituminous felt.[22] These coverings are applied over a structure which can be of timber, reinforced concrete or steel. In housing, steel is rare and, in the main, concrete roofs are limited to nontraditional construction or to blocks of flats rather than houses.

Flat roofs are today required to have insulation, properly designed and constructed to minimise heat loss; vapour barriers to minimise the risk of moisture condensing within the structure; and ventilation to release any entrapped water vapour.

Floors[23]

Above ground level[24]

Floors have consistently been made of timber, except in very recent construction where sound reduction and fire separation have made other materials more viable.

The floor structure consists of joists spanning between load-bearing walls. The direction of span will usually be the shortest distance. Where boards are visible, the joists run at 90° to the boards. As with the roof, loads are carried not just by masonry walls but also by internal partitions. In some construction – especially in 1930s semi-detached two-storey suburban housing – floors span the shortest distance in each room, thus requiring all walls to be load-bearing.

Timber ground floors[25]

These are of similar design to the upper floors. Instead of being supported on the external and other load-bearing walls, they are supported on small walls in the space below the floor. These walls, known as sleeper walls, are built off the ground or off a blinding of concrete and are surmounted by timber plates on which the joists sit.

Solid ground floors[26]

Where excavation below the lowest floor was not feasible, or too expensive, solid floors were used. These were commonly of stone slabs with joints filled with mortar until after the 1874 Act (see 'Damp-proofing' below). Poured concrete became the common material, laid on a bed of hardcore and finished with quarry tiles or a cement screed laid on to a damp-proof membrane.

Replacement floors where original timber floors have failed are commonly constructed of concrete where the depth for ventilation below the floor is inadequate.

Boards and sheets[27]

Until about 1960, flooring was usually of boards laid across and nailed down to the joists. Thicknesses varied and from the late interwar period to the early 1950s, tongue and groove boarding was common, intended to minimise warping and draughts. In the last 50 to 55 years, sheet materials – particularly chipboard – have become commonplace.

Damp-proofing[28]

Damp-proofing has been a requirement for parts of houses since 1875.[29] Dampness in this context is restricted to moisture entering the building from the enclosing or supporting ground. It does not include other penetrating or condensation-caused dampness, which is discussed in Chapter 3.

Walls[30]

Physical barriers in walls are installed to prevent ground moisture rising up to the level of occupation, usually the ground floor. The earliest materials were over-burned bricks, and these were superseded by slate, tiles, water-resistant engineering bricks, bituminous compounds, asphalt, metals and plastics. From 1874, any residential building was constructed with the intention that it should provide a barrier to ground moisture. Whether it was built with one or not is a matter of fact, but it should be presumed that it was unless shown otherwise.

Ground floors[31]

The lowest floor also required a barrier against ground moisture. In timber floors, this was by means of a damp-proof course below the timber plates or within the sleeper walls.

In solid floors,[32] this could be the stone slabs themselves or a quarry tile finish. In concrete floors, a membrane of bitumen, asphalt or, latterly, plastic was laid below the screed or floor finish.

Basements

Where the floor level is below external ground level, damp-proofing must include a form of tanking to be effective. Tanking is a vertical barrier against penetration of dampness, linked to the horizontal barriers in the floor and walls to form a watertight envelope.

Windows[33]

Sliding sashes[34]

Until the 1920s, vertical sliding sashes were the most common type of window for dwellings. They comprise a timber frame with cavities in which weights are suspended on cords to counterbalance the weight of the glazed timber sash, which moves up and down. The vertical axis is longer than the horizontal.

On more modern sliding sash windows, spring balances are used to replace cords. Although cheaper, they fail more frequently and require adjustment.

Casements[35]

These are either of timber or steel (a common trade name being Crittall). Their overall shape is usually rectangular with the horizontal axis being longer. Opening sections are hinged and they have catches and stays to lock them in position.

Replacements

With an ageing housing stock, replacement windows must be expected. These may be of the same pattern as the original but are more commonly of a cheaper and more readily available type. For example, sliding sash windows are often replaced with top-hung casements[36] or louvres.[37] In the last 20 years, plastic windows have become more widespread, offering cheapness combined with durability and a promise of low maintenance costs.

NONTRADITIONAL

Although not exclusively so, the majority of these dwellings are in the social housing sector, primarily in local authority developments. This was the sector that bore the pressure for production of high quantities of housing both between the wars and after the Second World War. It was the sector with money available in return for quantity and short-term solutions. The housing was never given a lifespan beyond 60 years, and commonly 30 years has been translated unofficially as 'in perpetuity'. Except in rare instances, social housing blocks of flats were of traditional construction after 1945.

The theories of architects and social engineers throughout Europe in the 1930s had resulted in the use of high-rise structures for commercial buildings. The structural qualities of steel and concrete enabled the engineers to design and build these multistorey buildings. It only required the architect to design an aesthetically acceptable facade and a single standard floor layout to produce a product offering simplicity, predictability and ease of repetition.

Steel and reinforced concrete frame

These were developed mainly in the interwar and immediate post-war years. The characteristics are a factory-made appearance without the dimensions or proportions of brick buildings. Large panels are often used to clad the exterior, and lightweight roof coverings are fitted on more slender structural components.

Large panel systems

The 1950s urge to build coupled with the shortage of skilled labour allowed the fulfilling of the long-held dream of builders to apply mass production factory techniques to the construction industry. This could be best achieved through prefabrication and limiting of site work to assembly. The large panel systems, marketed by different contractors under their various patent names, were the epitome of factory production in this period.

These systems were used for both low-rise and high-rise blocks, and although considerably out of favour, they still exist within the nation's housing stock.

No-fines concrete

No-fines concrete was used for housing in the 1950s and 1960s. The concrete is, as its title implies, without fines – i.e. sand and fine aggregate. The benefits of this are greater thermal insulation and speed of construction.

The outer face of the wall is rendered, and it is only the subtle regularity and absence of any sign of brick or block beneath the render that gives a visual clue to its construction.

Modern timber frame

The development of modern timber-framed construction began in the 1960s, and the construction of significant numbers of dwellings commenced in the mid-1970s. The structural part of the walls is a timber frame, treated against fungal and insect attack and protected by a waterproofing membrane. The exterior is often clad with a brick skin, in much the same way as the traditional cavity wall, with steel ties anchored into the timber structure. The external appearance is therefore of traditional construction, but (as seen in Chapter 3) very different defects can arise.

The telltale signs of timber-framed construction are the overall thickness of the walls: usually that of a traditional cavity wall (275mm/11") but with additional hollow plasterboard linings to the inside face.

Notes

1. Appendix I – Fig. 1.1 to 1.3
2. Appendix I – Fig. 1.1.
3. Appendix I – Fig. 1.2.
4. Appendix I – Fig. 1.3.
5. Appendix I – Fig. 2.1 to 2.2
6. Appendix I – Fig. 2.1.
7. Appendix I – Fig. 2.2.
8. Appendix I – Fig. 3.1 to 3.5
9. Appendix I – Fig. 3.1.

10 Appendix I – Fig. 3.2.
11 Appendix I – Fig. 3.3.
12 Appendix I – Fig. 3.4.
13 Appendix I – Fig. 3.5.
14 Appendix I – Fig. 4.1.
15 Appendix I – Fig. 4.2.
16 Appendix I – Fig. 4.2.
17 Appendix I – Fig. 4.1 to 4.2
18 Appendix I – Fig. 4.1 to 4.2
19 Appendix I – Fig. 4.1 to 4.2
20 Appendix I – Fig. 5.1 to 5.2
21 Appendix I – Fig. 5.1.
22 Appendix I – Fig. 5.2.
23 Appendix I – Fig. 6.1 to 6.2
24 Appendix I – Fig. 6.1.
25 Appendix I – Fig. 6.2.
26 Appendix I – Fig. 6.3.
27 Appendix I – Fig. 6.2.
28 Appendix I – Fig. 1.1 to 1.3, Fig. 6.2 and Fig. 6.3.
29 See Public Health Act 1875, section 157.
30 Appendix I – Fig. 1.1 to 1.3
31 Appendix I – Fig. 6.1 to 6.2
32 Appendix I – Fig. 6.3.
33 Appendix I – Fig. 7.1 to 7.6
34 Appendix I – Fig. 7.1.
35 Appendix I – Fig. 7.2.
36 Appendix I – Fig. 7.2.
37 Appendix I – Fig. 7.5.

Chapter 2

Materials of construction

Materials used for buildings may often appear to be inert. However, we have yet to develop materials for common use which are not adversely affected by the earth's natural environment. This chapter examines the nature of primary materials from which we continue to build houses.

If you do not know what a house is made of, you will not understand the nature or significance of a defect, or how a particular part of the building contributes to the overall performance of the dwelling as a shelter.

It is suggested that one of the marks of modern society, and perhaps of civilisation, is that we have sophisticated organisations and solutions to common human problems. We have evolved a system of construction intended to provide healthy buildings within which we work and live, shielded from the elements. This basic need to be protected from natural phenomena and risks has been recognised as a primary demand with the provision of a home being elevated almost to a basic human right, alongside liberty and freedom of expression. The question of whether a building can perform the function for which it was intended is always a relevant consideration.

A building is like an onion: one layer covering another. This chapter examines what materials are used in buildings and how things are put together as well as what the enemies of those materials and elements can do.

Appendix II sets out in more detail the analysis of defects from their initial appearance through to the identification of their primary causes.

Timber

Probably the most commonly found material in residential buildings, timber has been the basic material for construction since humans relinquished the natural shelter of caves.

Timber comes from trees. It is classified under two headings: hardwood and softwood. These groups are named not for the inherent feel of the timber, although in some cases that follows the name, but because of the structure of the wood. Pine – quite hard and durable – is a softwood, but Balsa – malleable, soft and compressible – is a hardwood.

Hardwoods take longer to grow and generally are not capable of replacement at anything like the rate at which they are used. Softwoods tend to be fast-growing and renewable. With the realisation that the earth's resources are running out, builders have been steered away from hardwoods except for particular uses where no alternative is suitable.

Timber is seasoned (i.e. dried) to a low moisture content prior to use. The lowering of the moisture content reduces the risk of warping and shrinkage once it is installed in a dwelling and the heating is turned on. It also reduces the moisture content to a level that is insufficient to promote or sustain fungal growth.

In domestic buildings, softwoods predominate. The nature and quality of the timber varies. Older buildings, particularly of the mid-Victorian period, used well-seasoned softwoods, durable and resistant to fungal or insect attacks (see Chapter 3). As building production increased in the later Victorian era and at various times since, the demand for good, well-seasoned timber has outstripped supply. To satisfy demand, younger or less durable softwoods have been used.

After years of problems with fungi and insects, the building industry accepted that poorer quality timber could only be used if it were treated with chemicals to resist more effectively the advances of those wishing to devour it. Today, almost all softwoods used in buildings are pretreated with chemicals to hinder the ravages of fungi and insects.

Two qualities of timber are used in construction. The cruder and non-planed timber is used for the structure, the floors, the roofs and the parts that are not seen when completed. This does not permit the builder to use a lower grade or strength of timber but refers merely to that which is not smoothed to an aesthetically acceptable finish.

The planed timber, shaved to a smooth surface ready for staining or painting, is used for the visible joinery in the building, ranging from floorboards to furniture.

The excess of demand over supply, with consequent increases in timber prices, also led to the search for new materials using otherwise rejected chippings, offcuts, etc. Composite, factory-made timber materials are now commonly found in buildings.

The most common is probably particle board (the generic term for chipboard, laminboard and myriad others). These boards are manufactured with offcuts, shavings and otherwise rejected pieces of sound timber, mixed and bonded together with resin to form a dense and solid mass. The density of the boards varies from highly compressed hardboard to soft insulation board.

These boards are used for flooring, kitchen units, shelving, doors and mouldings. Although apparently different to natural timber, they have the same properties and vulnerabilities to fungi and insects.

Brick

In the UK, we have large deposits of clay. Clay, when baked, forms a hard, dense and stable material that can resist the effects of rain and temperature and humidity changes. Since Roman times, bricks – initially for the better-quality buildings and eventually for all buildings – have been of baked clay. Modern bricks are manufactured to regular sizes and achieve a remarkable degree of standardisation in quality, thanks primarily to the improvement of oven-baking techniques.

Clay bricks are at their driest when they leave the kiln. From that point, they absorb moisture, although the extent of absorption diminishes with time. As the moisture content varies so does their size, causing movement within the structure.

In more recent years, other materials such as calcium silicate have been used for bricks. Unlike clay bricks, movement in these materials is primarily generated by drying rather than wetting. Therefore, their use in the ground presents no problems; but above ground, severe movement can occur.

In the nineteenth century, it was still necessary to have a skilled assessor of bricks to select those which were of better quality and to allocate inferior bricks to a suitable task. Today, this selection is done not from each batch but from the catalogue.

Because of the regularity in size and shape of bricks, bricklaying has become cheaper and faster. Today, notwithstanding the increasing use of other structural materials, brick is still holding its own as a major constituent of domestic buildings.

Bricks vary in density and strength. Some are extremely dense and almost impervious. These can be used as a damp-proof course or for below-ground work where dampness cannot be avoided. Above ground, bricks that are less dense, which have varying degrees of porosity, are used.

High-density, impervious bricks are expensive to produce; lower density bricks are far more common. Buildings are designed to accommodate this, just as cavity walls assume the possibility of water penetrating the outer skin of brickwork.

In most cities, older buildings are built of 225mm (9") brickwork (i.e. one brick thick). If the walls are in an exposed position or if there is a defect such as a leaking rainwater pipe allowing water to saturate the exterior, water penetration to the interior is likely. Thicker solid walls, denser bricks and the use of cavity construction are all factors that improve the walls' resistance to such water penetration.

Mortar and pointing

When built as a wall, bricks are bedded in mortar. This is a composition of cement, lime and sand in varying proportions. The mix, with water, is designed to hold the bricks together yet allow some flexibility. More modern mortar mixes are often hard. Whereas some structural movement can be accommodated with older buildings, every slight distortion in modern structures can be marked by a crack.

When pointing is missing, soft or perished, water can more readily enter the top surface of the brick. Combined with frost action, this leads to breakage of the brick and increases the possibility of water penetrating to the interior.

Render

Cement mixes are often applied to the outside of buildings. In some cases, this is for aesthetic reasons only, but more usually it is to provide a weatherproofing function. The mix, like mortar, is of cement, lime and sand and should provide a flexible surface. Again in modern times, hard, inflexible render mixes have been used. Crazing of render with minor non-structural cracks occurs with less flexible mixes or where the render is stronger than the brickwork on which it is applied. This crazing allows water to penetrate, and coupled with frost action, this causes deterioration of the brickwork often before the render itself manifests any serious failure.

Insulation

Modern buildings are required by the Building Regulations to achieve a minimum insulation value. For cavity walls, this is usually achieved by the insertion of insulation in the cavity. The materials commonly used are rock fibre (fibreglass), polyurethane foam and urea-formaldehyde foam.

Urea-formaldehyde foam has achieved some notoriety for discharging gases into dwellings. The chemical reaction of the two agents that make up the foam releases formaldehyde gas. The cavity must be well ventilated at its head so that this gas escapes.

Other risks with insulation in cavity walls arise from bridging of the cavity, allowing rainwater to pass from the exterior to the interior.

With solid walls, internal or external insulation is applied over the brick. The increased thermal efficiency will only be risk-free if any such insulation takes full account of moisture passing through the wall.

Rock

Rocks, either cut or crushed, supply a main source of building material. When used as a prime material without chemical conversion, this is commonly termed 'stone' (see the later section on stone).

Concrete

The chemical conversion of crushed stone into cement was perfected in the nineteenth century. Today there are a multitude of different cements which are used to make concrete. In all cases, the cement is mixed with sand and stones (together called aggregate) and water. The mixing provides both a physical blend and a chemical change which causes the cement to heat and cure into a hard and durable material.

In most domestic buildings, ordinary Portland cement is the main constituent of concrete. It develops its strength over a week or two. It has little resistance to acids or sulphates and may not always be used below ground where sulphate-resistant concrete may be required.

Other cement mixes achieve more rapid hardening or are especially resistant to acid or sulphate attack.

High alumina cement was rarely used prior to the 1950s. Its attraction was that it developed very high strength within 24 hours, thus speeding up the building process. It was particularly used in factory manufacture of prefabricated beams. After placing, high alumina cement concrete undergoes a chemical change known as conversion. The conversion results in a reduction in strength over five or ten years. When converted, the concrete is vulnerable to acid, sulphate and alkali attack. After ten years, some slight increase in strength occurs; but in warm, moist situations, further chemical action is possible depending on the aggregates used.

Plaster

The principal reason for applying plaster to wall surfaces is to conceal the unavoidable irregular surfaces of the structure itself. Plastering achieves this by filling the gaps and levelling out the surface. It also adds to the thermal efficiency of the wall or ceiling and to the sound reduction.

Where a wall has been damp, for example, where affected by rising damp, specialist plasters are applied to resist the passage of residual salts to the interior that would damage the decorations.

Stone

Stone is the term commonly applied to rock which is not chemically converted or otherwise restructured. It is rock that is cut from the natural source and split and shaped (dressed) and then used in its natural state as a building material.

Stone has been used for thousands of years as a building material. Although thought of as inert, stone may contain organic material and varies from very dense and strong (e.g. granite) to soft and porous (e.g. sandstone). The durability varies significantly from one quarry site to another.

Slate

Slate is also a stone and has been used for many years as a roofing material and as a damp-proofing medium. Due to its high cost and finite resources, imitation slates have been manufactured. Until the mid-1970s, these were primarily of asbestos cement, but today these are made of cement and mineral fibres.

Deterioration is generally slow and is aggravated by air pollution. Natural slate should have a life expectancy of between 50 and 80 years and imitation slate, perhaps 40 to 70 years. More likely than slow deterioration due to age is breakage of the slates by impact or corrosion of the metal fixings. With old slate roofs, iron nails fixed through the slates will have corroded well within the lifespan of the slate, causing the slates to slip. This is sometimes known as nail sickness. More modern roofs should have fixings of non-ferrous metals that are less affected by corrosion.

Tiles

Tiles for roof covering are made from clay and from concrete. The latter is cheaper and nowadays is the primary material in use. Tile profiles fall into two categories: plain and interlocking. The former simply lay one on top of another, like slates. Interlocking tiles are keyed into one another on either side.

Glass

Glass also originates from rock. It is manufactured from ground rock (soda-lime-silica). The most common type of glass is flat glass. This type is manufactured as clear float glass, patterned glass and wired glass. Tints and other surface treatments are applied to float glass for special applications, such as solar-controlled glazing.

Glass has very little thermal capacity, but it does have a high resistance to moisture. Because of this, windows are particular targets of condensation. The glass is at virtually the same temperature as the outside air and the surface is an impermeable vapour barrier. Run-off from window condensation is a common cause of defects to window frames of both metal and timber.

Since June 1992, there have been statutory requirements in the use of glass to extend the need to use safety glass for areas of risk. In particular, these are glazed doors, low-level glazing, side panels of doors and bathing screens. Many older properties have glazing which does not comply with these requirements. When reglazing, the present standards should be achieved wherever possible in the interests of safety.

Safety glazing is marked by small labels/engravings in a corner of the pane. Occasionally, glass that is vulnerable to impact may be coated with a protective plastic sheet.

Where glass forms part of a fire separating wall (e.g. between rooms) the glass is required to be wired or to have other special fire resistance. The size of panes is restricted so that the fire integrity of the wall is not compromised.

Metal

Metals are used for external protection (e.g. roofs and flashings), for structural supports, and for pipes, gutters, cables, etc. External protection is commonly of lead or zinc and is occasionally of copper. All of these materials are found as coverings of roofs and as flashings at junctions between roofs and surrounding walls.

Metals have a high propensity to move with temperature variations. Correct installation of metals therefore requires allowance for such movement. The construction of movement joints and the separation of large areas into bays that can move independently of each other are essential. Rigid joints, such as soldered joints in zinc, have a tendency to fail or cause buckling and failure of the metal itself if there is insufficient allowance for thermal movement.

Structural metalwork (e.g. steel beams) is factory made and assembled on-site. Joints are formed either by welding or by bolted or riveted connections. For protection against both corrosion and fire, structural steel is often encased in concrete. Where the steel does not require corrosion protection (e.g. over an opening in an internal wall), fire protection can be achieved by plaster or other fire-resistant cladding.

Asphalt, felt and other bituminous materials

Asphalt is used as a covering to flat roofs or shallow pitched roofs. It is applied in heated form as a liquid, and once cooled, it forms an impervious surface. Sand is rubbed into the surface to minimise the effect of the surface being coated with a layer of its bitumen constituent, which tends to craze. Solar reflective paints are often applied to asphalt to minimise temperature variations and oxidisation due to ultraviolet radiation.

Roofing felt is a bituminous material, bound together with mineral or synthetic fibres to form a sheet. It comes in rolls and is laid in either cold or hot compound over the roof structure. Although today some roofing felts are designed as single-layer coverings, the majority are laid in three layers, bedded in hot bitumen to form a single composite covering.

As with asphalt, solar reflection assists in preserving the integrity of the roof covering and this is achieved either by chippings, a top layer of mineralised felt (i.e. small chippings embedded in felt) or solar reflective paint.

For both of these materials, which are laid as impervious coverings, release of entrapped moisture or of vapour generated from below should be provided by pressure release vents. These look rather like plastic mushrooms over a roof surface.

Both asphalt and felt are used for damp-proof courses and generally perform well. However, as both become brittle with age, they are susceptible to failure where any structural movement occurs.

Medium Density Fibreboard

Medium Density Fibreboard (MDF) is an engineered material that can be used for a variety of purposes and costs less than wood. It is made by compressing wood fibres with a resin that acts as binder. The main component of MDF is softwood, but there can be varying amounts of temperate hardwood if this is locally available to the manufacturer – found especially in MDF manufactured outside the UK and Ireland. Different binders are used, depending on the intended end use: urea-formaldehyde is used for dry environments; melamine urea-formaldehyde, phenolic resins and polymeric diphenylmethane diisocyanate (PMDI) are used where moisture resistance is important. PMDI binder is not formaldehyde-based.

There are potential hazards associated with the machining or sanding of MDF due to the presence of formaldehyde, which is classified as a carcinogen (as well as the presence of hardwood and softwood dust). However, levels of formaldehyde in MDF made in the EU are considered to be within safe limits.

Innovative materials (and off-site construction), allowing for greater levels of insulation and water control, are becoming more common, used primarily in new construction. However, as we see later, some of these methods can result in defects in use. This is often because, unlike for a computer or a car, we do not provide a layperson's guide to the use and occupation of housing but, rather, expect everyone to understand the need to achieve balanced occupation (e.g. ventilation/heating/insulation) and to apply appropriate heating.

Chapter 3

The enemies of healthy buildings

If the materials from which we constructed our buildings were not largely inert then the adverse effects of the elements would be unlimited. Buildings would be eroded by rain, blown down by wind, overloaded by snow, overheated by the sun. These phenomena do take their toll on our buildings, but in general, building engineering has reduced these risks to a minimum.

Housing managers are less likely to be faced with houses washed away by floods or in a state of near collapse as a result of excessive winds – particularly in the UK – than they are to be faced with more insidious degradation of buildings. Nonetheless, over recent years and with the effects of climate change, such extreme weather is occurring more often.

Flooding requires immediate protection to restrict further water ingress. Pumping out of surplus water is required, followed by drying out of the dwelling. Drying can take a long time, as to do this too speedily may cause further damage as materials readjust to their dry condition. Following any such drying out, there is residual dampness that poses a clear risk to services such as electrics, drainage and water supply, all of which may require renewal.

Wind damage is often more visible where sections of the building have been blown away (e.g. roof tiles); though there may also be damage to sections that are still standing but which may have been weakened (e.g. chimneys).

Of all of the defects likely to be encountered, dampness is the most common and potentially the most harmful, both to the structure and to the occupants.

Water

Water can be an enemy of buildings in three ways.

First, water affects the structural stability of buildings by varying the supporting qualities of the subsoil. Foundations are affected by water. When buildings are constructed, foundations are designed so that – except in extreme weather conditions – they will be unaffected by temperature or moisture content. Older houses have shallow foundations and are more susceptible to seasonal movement.

Second, and perhaps more commonly, water gives rise to unwelcome life. The source of life out in the garden is also the source of life within the dwelling. When water comes into contact with any organic material it provokes life. It also attracts and sustains life.

Third, when an inorganic material – such as brick, plaster or concrete – becomes wet, chemical changes occur. Carbon dioxide and sulphur dioxide are present in the air. When combined with water, these can convert to harmful acids that attack stone and concrete as well as brickwork.

The identification of the source(s) of the water is essential to specify remedies, but whatever the source, its effect is broadly the same. This is often forgotten, giving rise to statements such as, 'It's not damp – it is only condensation'.

Moisture passes through the structure by capillary action and direct flow. If allowed, it will penetrate brickwork and concrete; this can only be halted by an impervious barrier.

Damage from water does not occur only when there is a flood. Continuous and even intermittent dampness also causes defects, and in the UK, this is the most common source of building defects.

Dampness from the ground

Rising damp

Even in a drought, the ground takes a long time to dry out. The water level and content in the soil may vary and, as seen elsewhere, this may cause structural problems, but the soil will still be damp and capable of transmitting that dampness to the building above.

Rising damp was the primary source of dampness in buildings in the nineteenth century. It was this defect that succeeding Public Health Acts and Building Acts sought to remedy along with insanitary drainage and washing facilities. It has still not been cured over a century later.

Rising damp can be confused with other sources of dampness. It has become the common name for all dampness originating out of the soil. However, its true meaning is specifically the moisture rising vertically within a wall or through a floor. This distinction is important when identifying the source of and liability for a defect.

In modern housing, impervious barriers to rising damp are unlikely to fail if correctly installed. Failure of older damp-proof courses such as slate, bitumen or asphalt is likely due to the deterioration of the material. Where structural movement occurs, this may provoke rising damp due to fracture of the membrane.

Remedial treatments such as silicone injection can be successful. However, the design and installation of these remedial actions depend not just on the treatment itself but on a whole series of other factors. It is the failure to deal with these other factors that is the primary cause of recurrent rising damp.

The narrow definition of rising damp is particularly relevant where remedial works have been carried out but dampness recurs. In most cases, an injection damp-proof course is covered by a guarantee. However, the guarantee rarely covers the recurrence of damp. What it warrants is that the damp-proof course injection will not fail. Therefore, when dampness recurs but the injected barrier is still working, there can be no claim on the damp-proof installation.

It is with this in mind that other sources of dampness, associated with rising damp but not actually rising through the structure, must be considered.

Lateral penetration

Lateral penetration does not just occur horizontally. Penetration of moisture through a wall or floor may occur where the damp-proofing works but is being bypassed.

Where the external ground level is above or close to the damp-proof course, moisture passes across or around the impervious layer.

Rooms below ground level are particularly vulnerable to lateral penetration as the external walls are earth retaining. The walls of older properties are protected externally with materials similar to those used for a horizontal damp-proof course.

Modern basements are constructed with an impervious membrane on the exterior of the wall linked through to the floor to provide a complete tank, rather like a swimming pool but with the water only on the outside. The floor and walls support the membrane, and the load from the ground holds it in a sandwich between the structure and the soil.

In old basements, or in any buildings where the floor is below external ground level, remedial works can only be successful if they replicate this modern construction. This requires neither demolition and rebuilding nor excavation around the perimeter. It does require the sealing of the walls to the floor and the application of material to the walls, which withstands the hydrostatic pressure from the groundwater. This remedial treatment, known as 'tanking', is common but nonetheless problematic.

Remedial damp-proofing

Remedial treatments to older buildings are generally carried out by specialists. The level of expertise does vary considerably, however, and the acceptance of responsibility for recurrence of dampness can be very difficult to achieve. Virtually all remedial treatments work well in the laboratory. The problems start when they are applied to real buildings.

The repair of dampness from the ground depends totally for its success on the condition of the base that is to be treated. Old walls often contain voids into which the injected silicone can flow, thus not forming a complete horizontal barrier in the masonry. Broken brickwork in walls does not accept a silicone injection; neither do perished and friable bricks provide a stable base on to which cement tanking can be applied. Floors must be stable and of adequate strength to accept a damp-proof membrane. If the base on which the repair is applied fails then the repair itself will fail.

Bridging

Solid floors at or below ground level should contain a damp-proof membrane to resist the moisture from the ground. To provide a comprehensive barrier to dampness, this membrane must link to the damp-proof course treatment of the enclosing walls.

In cases of recurrent dampness in older properties, I have found that the most common failure is not with the injected damp-proof course nor with the floor membrane, but the link between the two. The reason for this is that although the wall is treated by a specialist, the floor membrane is laid by largely unskilled labour as part of the placing of the concrete forming the floor structure. A little extra care at this stage would save thousands, if not millions, of pounds across the UK.

Where the membrane has been covered by the concrete floor and not turned up at the wall abutments, the junction between the concrete and the wall forms a ready route for moisture.

Similarly, plaster internally or render externally which covers the damp-proof course allows bridging; i.e. capillary action taking moisture past the membrane.

Salts and residual dampness

Moisture from the ground contains salts. It is the presence of these salts that assists in the identification of the source(s) of dampness (see Appendix II). When repairs are carried out, the salts remain in the masonry. Unless the plaster internal to the building contains an additive to act as a salt inhibitor, these salts will migrate to the interior and cause staining and damage to decorations. The salts will also contaminate the plaster and absorb moisture from the atmosphere, resulting in recurrent dampness.

Treatment of a damp wall does not make it dry immediately. As a rough guide, the wall dries out at the rate of one month for each 25mm (1") of thickness. On this basis, a solid wall the thickness of one brick takes nine months to dry out.

Where possible, the building should be left unoccupied and gently dried out. In the real world, this is rarely possible. Replastering after a remedial damp-proof course has been inserted has to ensure that the migration of this residual dampness to the interior is resisted.

This in itself slows the drying process as the moisture then primarily dries externally. However, by applying the correct mix of plaster, some moisture discharge is possible internally without causing undue discomfort or damage. It is for this reason that following damp-proofing work, only waterborne, not vinyl, emulsion paint should be applied to the walls until the drying out is completed. Under no circumstances should impermeable wall coverings be applied for at least 12 months after treatment.

Dampness from the ground can therefore be due to:

- absence of an effective barrier to rising damp; and/or
- absence of an effective link between the wall damp-proof course and the damp-proof membrane to the floor; and/or
- the bridging of the wall damp-proof course by render externally or plaster internally; and/or
- the migration of salts from the treated wall through an inappropriate plaster mix to the interior; and/or
- residual moisture in the structure.

Appendix II assists you in narrowing down these possibilities, but do not be tempted to eliminate any of them until you are 100 per cent sure that it cannot apply.

Above ground

'Above ground' in this context means above the damp-proofing layer. It therefore includes the lowest floor.

In buildings today, pipework is rarely run in a concrete floor. This is because the pipes are generally of copper, which corrodes if unprotected in cement. The corrosion perforates the copper, leading to leaks. Unfortunately, the majority of buildings built up to about 1980 have pipes buried in the floors. The practice has still not been eliminated, but it is less prevalent.

When pipes leak, water is discharged and causes damage. Weeping pipes, connections and fittings can cause substantial damage over a period of time, even though the leakage may be intermittent or at a low rate.

Plumbers are increasingly asked to fit washing machines and dishwashers that are built-in or partly concealed by worktops, cupboards, etc. As the pipework is concealed, these appliances may need checking to find the source of a leak.

From above

When water enters at one point, it does not usually flow directly down to the interior. Below the external covering is a whole series of layers, each of which absorbs the water before it passes to the next layer down and eventually to the interior. Water finds the easiest route, and even if it has penetrated directly to the upper surface of a ceiling, it will look for cracks and joints in the plaster to pass through to the room below.

A very small failure in the covering can give rise to significant water penetration. Once water has entered in this way, it can have the same effects of vapour build-up and deterioration of the structure as water from any other source.

Cavity construction for walls

Cavity walls have a special requirement for damp-proofing. It is part of their design that water can enter the outer layer but cannot penetrate to the interior. Around openings and wherever the cavity is bridged – by a floor beam, for example – a damp-proof membrane has to be installed.

As with so many parts of the building process, quantity gives precedence to quality, and these damp-proofing measures have often been omitted or installed inadequately. Remedial works can be expensive and disruptive.

Where insulation is installed in the cavity (see Chapter 2), the choice of materials and the installation must ensure that a bridge is not created allowing external moisture to penetrate to the inner surface. The materials used in new work are fixed so that an air gap remains between the outer face of the insulation and the inner face of the external skin. For remedial insulation systems, the use of moisture-resistant and largely impervious materials minimises the risk of water transmission.[1]

From inside

Condensation

Air contains water held in vapour form. The hotter the air, the more water vapour it can hold. The quantity of vapour in the air is measured as relative humidity. There is no inherent problem with this moisture until the air meets a colder and impervious surface. The air is cooled and can hold less water. The excess water condenses out on to the colder and impervious surface and forms water droplets. The water from this source does not contain as much salts as water from the ground, but it can cause an equal amount of damage to buildings.

The balance between heating, ventilation and insulation needs to be right to overcome the problem of condensation. It is affected by moisture generation, but basic failings have to be present in the building for any usage – except that which is extremely unreasonable – to cause severe problems. The adjustment of any of the three factors modifies, and may ameliorate, the conditions, but it is only the balancing of all three that can cure the problem.

Relative humidity levels of around 50 to 60 per cent should maintain an adequate comfort level without excessive dryness and should not produce excess dampness. However, if humidity levels increase to 70 per cent or more then condensation is common and this sustains and generates the moulds and insect infestations referred to later in this chapter.

Interstitial condensation

This is condensation occurring within part of the building structure, commonly in roofs. Water vapour entering a flat roof structure struggles to escape or else condenses on the underside of a colder, impervious layer (usually the underside of the roof covering). This has two effects. First, it causes the structure of the roof to become damp, with all the consequences following (as described). Second, build-up of vapour in the roof exposes any weakness in adhesion of the roof covering, causing it to bubble upward.

From construction

Vast quantities of water are used in construction. Most of the materials used in buildings either contain water or are mixed using water. All of this moisture has to dry out and be removed from the building.

Sulphates

Brickwork is vulnerable to chemical attack. Sulphates from rainwater or from constituent parts of the building react with Portland cement (a primary constituent of mortar). This reaction causes the expansion of the sulphate, in turn causing expansion of the brickwork. Where cavity brickwork is used or where solid walls are constructed of bricks of different quality for the inside and outside, this expansion can cause distortion of the wall.

Gases in flues can also often cause sulphate attack to the chimney, at which level the flue is not lined and the gases condense out on the inside face of the bricks.

Fungi

Fungal decay attacks all organic materials, not just timber. Fungi are plants, and they grow and reproduce. Their spores (seeds) are in the air and these germinate wherever there is an inviting host.

In buildings, we are concerned with those fungi which live off dead material. The identification of broad categories of fungal decay is necessary as the consequences of their actions and the repairs required vary widely.

Dry rot

Dry rot (*Serpula lacrymans*) is a fungus which has remarkable powers of survival. Thought to originate in the Himalayas, it is a possible by-product of Britain's imperial past.

As with other wood-rotting fungi, it requires a moisture content of dead timber in excess of 19 per cent to start germination. What distinguishes it from other fungi, however, is its ability to survive and grow even when that moisture level is reduced. Dry rot is able to survive in timber with a moisture content of 14 per cent. On the other hand, where the moisture content is 40 per cent or more, it is unlikely to grow.

This ability to spread in relatively dry conditions and to continue spreading after a water source is removed makes it the most difficult to treat. Treatment must extend to at least 1m beyond the last growth and include removal of all organic material. In older dwellings, this includes plaster that contains animal hair or other organic material as a binding agent.

Dry rot is able to travel through brickwork, finding passages in mortar joints and feeding off hair and other organic material. It attacks furniture and belongings, and can remain concealed until finally manifesting itself when it is too late.

When visible, its first characteristic is warping or wrinkling of the wood, often with the outer, exposed surface remaining intact while the material behind decays. The decaying material affected by dry rot has a cuboid cracking pattern, both along and across the grain of the timber. This cuboid pattern is also found with some wet rot fungi (see below); but in all cases, erring on the side of caution is recommended and where such patterns are found, dry rot must be suspected.

In advanced stages, and only in concealed areas, mycelium in white/cream sheets with tentacles is found on the surface of the affected material. When threatened, the fungus seeks to reproduce and creates a fruiting body. These can be really beautiful. They are flat, slightly puffy growths with white perimeters and rust-red centres. The red part is the spore bank from which the spores are spread by air movements to new feeding grounds.

Wet rot

Wet rot fungi, of which there are thousands if not millions, require high moisture levels to survive – usually in excess of 40 per cent. Once the water source has been cut off, the fungus dies.

The manifestation of fungal decay by wet rot has a common pattern: the distortion of the affected surface of the timber and disintegration of the material. In most cases, there is also a dark staining, which is a useful – but not 100 per cent certain – guide to distinguishing a wet rot attack from dry rot. The affected timber will still be damp if the attack is active.

Remedial treatment is therefore limited to replacement of unsound material and chemical treatment of the affected areas. The chemical treatment must be carried out. However, it is common for a wet rot attack to be cured by cutting off the water source only to find that dry rot takes over as soon as the moisture content declines.

Moulds

Many of the moulds found in dwellings are allergenic. They can give rise to allergic symptoms in the nose and throat. The allergic reaction of occupants will depend on the toxicity, buoyancy and concentration of the allergen.

Moulds require moisture (some more than 65 per cent relative humidity), moderate temperatures (13°C–15°C) and suitable food. The food source is commonly wallpaper paste but may also be clothing and belongings.

The most effective way to control mould spore allergens is to reduce moisture levels in the air. Increased ventilation can relieve these problems by reducing moisture levels in indoor air, but this has to be balanced against temperature levels and the ability to maintain heat levels.

Insects

Wood boring

Wood-boring insects attack dead wood. Their presence is identified by their flight holes, usually small in diameter. It is the weakening of the timber by these holes along with the cutting out of the internal passage by the insects for food that cause the damage.

Detection in advance of some damage is difficult if not impossible. The insects are minute, often not readily visible to the naked eye. The only manifestation of their presence is the flight hole after they have gone and a small pile of frass (digested and excreted wood dust) left around and below the flight hole.

A telltale sign of deathwatch beetle, which attacks oak, is the knocking sound it makes. It is unlikely that many housing managers will work with buildings using oak, and even less likely that they would be able to stay there in silence to hear the tapping sound. However, if buildings contain oak, an occupant's complaints about tapping sounds in the middle of the night should not be dismissed as irrelevant to disrepair.

These insects have a seasonal life cycle. The emerging insect lays eggs on the surface or within the wood. The egg germinates and the grub begins its life within the timber. The grub feeds off the wood and, after metamorphosis into a flying beetle, emerges by eating its way through to the exterior.

Some wood-boring insects, such as the deathwatch beetle, do not emerge every year; but the presence of flight holes, especially if frass is visible, indicates that the infestation is active. Remedial treatment is, therefore, required to prevent subsequent generations from boring more holes and causing damage. Treatment against common insects is now almost routine in rehabilitation work. In new buildings, pretreated timber is normally used.

Disease carrying

Cockroaches have developed a survival rate that has to be admired. They can move from one dwelling and contaminate another while the first is being treated, then reoccupy when safe to do so. For this reason, treatment of individual dwellings in a block of flats is unlikely to be successful. Whole-block treatments are the only hope. Many local authorities have rolling programmes of whole-block treatments, which require repetition at frequent intervals to overcome this type of infestation.

Ants can also occupy homes, feeding off foodstuffs. They prefer damp, dirty and unhygienic food stores. As with cockroaches, their eradication requires whole-building treatment.

Fleas (both human and animal) can usually be treated effectively, but their removal requires the source (e.g. a pet) to be treated also.

House mites are a major source of allergens. They are barely visible to the naked eye and are usually found in house dust and in bedding. The mite feeds on dead human skin. They require a high humidity level (not less than 45 per cent) and, therefore, damp conditions increase their population and activity.

Just unpleasant

Silverfish, wood-lice and other harmless insects also invade the sanctity of our homes. These are sometimes inconvenient and unpleasant but do not carry disease or provoke allergic reactions.

Metals

Metals corrode when affected by water, but this is not due only to the effect of the water. Metals are affected by electrolytic action – this occurs when two different metals are linked by a conducting medium such as water. In buildings, the most common metals to react together adversely are zinc and copper. Zinc is used not just in its own right but also as the galvanised coating to steel. Galvanised water pipes and water tanks joined to copper pipes will corrode.

Hazardous materials

Asbestos

Materials containing asbestos were used for various purposes in our buildings before it was banned in 1999, having been linked to lung diseases and cancer. The Health and Safety Executive (HSE) advises that it is still present in many buildings and the risk materialises when these materials are moved or if their condition has deteriorated; in these instances, asbestos fibres can be released and breathed in.[2] The material is used in various forms and there are three distinct types (see Table 3.1).

Its use today is almost non-existent, but it is the treatment of the material already present which has to be carefully considered. The current regulations – the Control of Asbestos Regulations 2012 – came into force on 6 April 2012, updating previous asbestos regulations. Under regulation 4, there is currently a duty to manage asbestos in *non-domestic* premises, though this also applies to some parts of domestic premises – for example, common parts of premises such as housing developments and blocks of flats. However, there are *no direct duties on landlords for individual houses or flats*.[3]

Table 3.1 Types of asbestos, their applications and removal

Asbestos-based materials	Uses and composition	Stripping out
Asbestos cement boarding	Commonly used as roofing material and for water tanks, flue pipes, etc. Contains about 10 to 15 per cent chrysotile and 85 to 90 per cent Portland cement mixed with water. Flue pipes may contain amosite.	Wet stripping may be permissible if there is no prospect of breakage or fracturing, but where this is not guaranteed then an enclosure should be provided.
Insulation board	Commonly used as fire-resisting board. Contains up to 40 per cent amosite or amosite mixed with chrysotile and 60 to 84 per cent calcium silicate.	Stripping can only be carried out using enclosures.
Rope and gaskets	Commonly used to seal flexible joints. Can contain over 90 per cent asbestos, usually chrysotile or crocidolite.	Stripping can only be carried out using enclosures.

Removal contractors

The regulations outline which types of work with asbestos must be carried out by a licensed contractor and where work with asbestos is not licensable. Notification must be given to the relevant authority for any work that is licensed and for some circumstances where it is not licensed. For instance, it is not necessary to have a licence to carry out work with asbestos cement, though the work is notifiable where materials have been damaged or disturbed.

Removal/encapsulation techniques have improved over the years with most, if not all, contractors being well versed in safety issues and protection of occupiers. Education and training has resulted in a greater awareness of the dangers. The HSE provides guidance sheets on different aspects of working with asbestos; and the Asbestos Removal Contractors Association is attempting to maintain high standards through provision of information and guidance as well as monitoring of members' performance.

Glass fibre

Glass fibre wool is used in buildings as an insulant, usually in the form of matting. It is found in lofts where it is laid over the ceiling or under the roof as well as being present in cavity and timber walls, all to reduce heat loss. It is also used for pipe and tank lagging and, in a denser form, as a sound insulation quilt.

Glass fibre wool has raised similar safety concerns to asbestos in the past but is not considered to be anything like the same hazard. Nevertheless, strands cause irritation to the skin and to breathing, and suitable precautions should be taken when handling this material.

Urea-formaldehyde foam

Urea-formaldehyde foam is used as an insulating material in cavity walls into which it can be pumped. When mixed, it releases formaldehyde. At low concentrations, this irritates the eyes and instils nausea and some breathing difficulties.

Radon gas

Radon gas is the result of the breakdown of uranium-238. The gas seeps out of the ground and can enter buildings. Increased insulation and restricted ventilation has resulted in a higher risk to occupants of dwellings. Concentrations vary throughout the UK, but where this is high, increased incidences of lung cancer have been reported.

Users

Buildings would operate perfectly well if no one occupied them. This may seem to be a ridiculous statement, but we cause no end of problems when we start to live in these structures. In particular, we cause dampness.

Moisture generation

We all generate moisture by breathing, by washing ourselves and our clothes, by cooking and by heating. For example, four people in a house for 12 hours generate 2.5kg of moisture. Cooking and food preparation can produce 3.7kg per day, floor mopping, 1.1kg, clothes drying, 12kg and clothes washing, 2kg. The construction and design of dwellings should overcome the generation of these relatively large amounts of water by balancing insulation, ventilation and heating.

The use of flueless gas heaters (e.g. those using Calor gas) and paraffin heaters creates further water vapour. For example, a paraffin heater discharges a litre of water vapour into the air for each litre of oil burned.

Sweating also generates moisture. While we expect sweating to occur at higher temperatures, it must be remembered that this is related to humidity levels as well as to temperature. For example, at a humidity level of 22 per cent, profuse sweating does not occur until $30°C$; but at a humidity of 60 per cent, it occurs at $20°C$ to $25°C$.

Animals

Disease carrying

Pets, the great love of the British, are a primary source of fleas in our homes.

Disposal of faeces from pets is also a real problem, especially in blocks of flats. The effects of dog faeces on health have been well publicised and proper hygienic cleansing of areas used for defecating by pets must be maintained.

Rats and mice are probably the most common invaders. Rodents carry many diseases mainly in the parasitic fleas which live on their bodies. They will, if cornered, attack humans and care must be taken in entering cellars or sewers where rats are likely to be present.

Refuse

We produce refuse which we fail to properly dispose of or even to seal up pending disposal. We flush our waste down the drains but how often do we clean those same drains?

Plants, trees and bushes

It is also worth looking at the effect of trees and bushes. Although we perceive plants as growing where we plant them in our gardens or window boxes, they have a life of their own. The seeds are in the air and will propagate wherever there is a source of food and water. They do not exclude parts of our buildings from this. In some instances, plant growth is a result of poor building maintenance. Where pointing to brickwork has been eroded, plants often take root. In other instances, plants can expand beyond the confines of the plant pot and, particularly with creepers, seek out walls over which to climb.

Plant-covered walls can be attractive to look at but, with few exceptions, they conceal the damage being caused to the fabric of the building. Plants need water and will obtain it from any source. The roots that are produced to search for and absorb moisture are capable of finding a way through minimal openings, even between bricks. As these roots develop they expand. The expansion widens the gaps between the building elements, allowing more water to enter and setting in train a cycle of deterioration.

Trees and bushes also have adverse effects on buildings, but this is primarily external. As shown below, the moisture in the soil, affected by drought or flood, can have a damaging effect on buildings. Trees and bushes need water to survive and in times of drought, their root growth extends to search out water; this along with their extraction of the limited moisture from the soil causes further damage. As a result, buildings have cracked during hot summers.

Temperature and climate

Drought

The main effect of drought is on the subsoil and on the foundations, but high temperature levels also cause unusual thermal movement of the building elements, putting unexpected strain on the joints and junctions.

In times of drought, subsoils such as clay shrink. The voids thus created allow for compression and slipping of the clay layers, which in turn removes support from foundations of buildings. The subsequent cracking of buildings is known as settlement.

As noted above, this drying out of the subsoil is hastened by trees and bushes extracting the remaining moisture. Often, therefore, serious settlement problems are a combination of changing subsoil conditions and root growth.

Frost

The main effect of frost is on the exposed surfaces of the building. Water expands when frozen. If water has entered the building fabric and then freezes, the expansion causes damage, splitting bricks and breaking concrete. The opening thus created lets in more moisture, continuing the cycle at an accelerated rate.

When tanks and pipes are not lagged, freezing opens joints and splits pipes. When the temperature rises, the water flows.

Snow

Snow imposes significantly increased loads on roofs. Modern roofs are designed for normal snow loads, but structural failure is possible in exceptional conditions. Snow can also pull slates and tiles down roof slopes.

Rainwater pipes blocked with snow are not able to carry away the water discharging from the roof when the thaw starts. There is a time delay between the warming of the roof and the thawing of the ice in the pipes. During this period, overflowing and leakage should be expected.

Sun

The sun discharges heat and radiation. The heat has a direct effect on buildings by causing expansion of materials, which in turn can cause fracturing of the less flexible elements. Ultraviolet radiation from the sun causes oxidisation of asphalt and breaks down surfaces that use this material. Temperature changes affect some materials more than others.

Underground threats

In areas where mining, quarrying or other extraction industries are (or were) found, there is a risk of subsoil movement. This occurs due to the falling of soil into the voids left after extraction and is known as subsidence. The effect on buildings is very similar to settlement, but the cause is distinct. The difference is important not just in the remedial works required but also in the securing of insurance and the possibility of recovery of costs of remedial work.

The effects of underground streams are often forgotten, particularly where there is a propensity to settlement (e.g. in clay subsoils). Watercourses erode subsoils, giving rise to subsidence. Water finds the easiest route, so when desiccation of the subsoil occurs, either through drought or water extraction, streams may well change course. This in turn can give rise to problems of flooding and erosion of subsoil in an otherwise unaffected location.

Nontraditional buildings

Buildings are unpredictable. Traditional buildings do not all fail in the same way, or at the same time. And with nontraditional buildings, failures depend on the same multitude of factors with similar scope for variation.

Prefabricated buildings

The Housing Defects Act 1984 designated as defective certain types of prefabricated dwellings which were known to be vulnerable to major structural deterioration, making them unsaleable. The Act defined the types and, through accompanying information papers, gave details of the construction and vulnerability of these types. These were all of reinforced concrete construction. Other prefabricated buildings were constructed in the interwar and immediate post-war years, including many steel-framed designs. British Iron and Steel Federation (BISF) houses were manufactured in substantial quantities and can be found throughout the UK. The designs and potential defects in these buildings can be found in Building Research Establishment (BRE) publications.

Large panel systems

These were used extensively in the UK during the 1960s housing boom. They were manufactured by individual companies and sold as a kit, assembled on-site. Although the designs worked beautifully on the drawing board, the designers had reckoned without the workforce. A combination of unskilled labour and a zeal for quantity rather than quality led inevitably to substantial construction defects. Perhaps the most notorious was Ronan Point which partially collapsed, but others have been more of a financial liability for their owners. Often the repair costs have been so high that demolition and rebuilding has been the only sensible solution.

The main sources of defects were the joints between panels. These were either not properly linked structurally or not adequately sealed against the weather.

The enemies of these buildings are in many ways the same as for traditional buildings, but their 'pack of cards' structure means that localised failures can have a devastating effect on the whole structure.

High-rises

Apart from the human and social costs of the 1960s adventure, high-rises pose special construction problems. These apply whether the structure is of steel, reinforced concrete or prefabricated construction and are primarily linked to the height and exposure of the building.

The propensity for water penetration is always governed by the exposure of a building. High-rises are clearly more exposed to wind and rain. The buildings are most vulnerable at construction joints and around openings. Windows originally installed were often of low quality, acceptable for low-rise buildings but not sufficiently sealed for high-rises.

Further defects to which high-rise blocks are vulnerable are the accumulation of common defects found in all buildings, which together have an increased effect. Ventilation to bathrooms and toilets is often via a shared vent stack, which may or may not be powered. If the stack becomes blocked or vents leading into it are permanently open, exhaust from a unit low down in the block will discharge to upper units. There is a real danger of fire spreading through the ducts from one unit to another. The installation of fire checks at each floor level has often been a casualty of the quantity versus quality battle.

Blockage of above-ground drainage can exert massive pressures on the pipework, causing spectacular and horrible explosions of waste and sewage. Vacating units for repairs can be an organisational nightmare.

Modern timber-framed buildings

Timber-framed buildings are quicker and cheaper to erect than those of conventional brick. They can also provide higher levels of insulation and comfort than masonry. However, the primary element of construction is timber and this is affected adversely by moisture. Vapour barriers – impervious membranes – are fitted within the structure to protect the timber from moisture from the exterior. Insulation is fitted to maintain the timber at a higher temperature, closer to that of the interior, so that condensation does not occur on the inside face of the vapour barrier. It is the failure of the detailing of this insulation and vapour barrier that is the most common source of moisture penetration.

Although water leakage and infiltration from other sources has a damaging effect on all buildings, this can cause major structural weakening for timber-framed buildings. Therefore, whenever dampness occurs, the likely effect on the timber structure must be considered.

Notes

1 See Appendix I – Fig. 1.2.
2 See Health and Safety Executive (2012) *Asbestos: The Survey Guide* (2nd edition), available at: www.hse.gov.uk/pubns/priced/hsg264.pdf
3 Further information on the duty to manage asbestos is available at www.hse.gov.uk/asbestos/duty.htm

Part 2

Inspecting disrepair

Chapter 4

Preparing for the inspection

The housing manager's role

This book is concerned not just with complaints by occupants but also with the need for housing managers to be aware of what defects may exist that are giving rise to the complaints. Housing management staff are often required to be alert to 'asset management', which involves keeping properties repaired for the sake of the occupants alongside protecting the landlords' investments and complying with their legal obligations. In this era of private finance, lenders may require that any stock on which a mortgage is secured is maintained in good condition to preserve its value.

Maintaining a property in good repair and solving problems as they arise is important for the prevention of legal disputes. In *Housing: Proportionate Dispute Resolution: An Issues Paper*, published in March 2006, the Law Commission looked at the role of courts and tribunals in disputes related to housing. The view was taken that a 'holistic approach' for 'proportionate' resolution of housing problems and disputes can be developed (in addition to a more coherent legal framework). This would focus on solving problems early on and, where possible, without the need for formal legal processes. The role of the housing manager, in terms of problem-solving and being a main point of contact for tenants, is key to this approach.

'Following the trail'

If the housing manager is to be effective in the first reaction to and diagnosis of defects then he or she must think about the underlying cause of the problem. This is not to say that every housing manager has to have the experience, training and skills of a qualified surveyor, but she/he should have sufficient common sense and knowledge of buildings to direct a train of thought to at least eliminate the more serious implications.

In terms of disrepair, when surveyors conduct inspections of properties, they are expected to 'follow the trail'. This term has arisen out of actions taken against surveyors who have allegedly failed to notice or to report on defects. The surveyor is expected not just to see what is visible but also to raise questions prompted by what is seen.

For example, where a timber ground floor slopes to one side, it is not sufficient for a surveyor carrying out an inspection for a prospective purchaser to merely report the slope. It is expected that the surveyor will also assess the likely cause of the sloping and advise on what risks are posed and the possible implications. Advice on further investigation or tests to refine the assessment is also required. After all, it is this very advice which the surveyor's skill and experience is being called upon to supply.

Whereas the further investigation and analysis of data to conclusively identify an underlying defect may be beyond the skills required of a housing

---

Box 4.1 Example

Some 40 years ago, a local councillor requested advice for a group of residents involved in a consultation exercise for an estate improvement scheme. The buildings were pre-1939 blocks of flats that had originally been fitted with solid fuel fires, but these had been replaced with gas fires during the 1960s. There was no heating to rooms other than the living room.

In relation to the consultation, the landlord (in this case the local authority) had based its assessment of priorities on the complaints received from tenants over the previous five to ten years. The most common complaint had been about kitchen units. These had been installed in the 1960s and were, without exception, of laminate-covered chipboard. They had disintegrated as chipboard does when it gets wet.

The landlord's presentation to the residents was done on the basis that the highest priority was to replace these units. The residents were arguing for higher quality units of solid timber rather than particle board, and the whole process was getting bogged down.

The authority had not consulted its professional advisers at this stage but had identified the cause of the problem with the units as dampness. It had failed to look beyond this to see where the dampness originated, which in this case was condensation caused largely by the inadequate heating in the dwellings. What was required were remedial works to minimise this condensation-caused dampness; but for some reason, no one had interpreted the original complaints from tenants in that way.

To replace the units without addressing the underlying cause would have been a waste of public money and a source of further dissatisfaction among the tenants.

manager, the identification of a trail to follow should be within their experience. Having identified that a trail or trails exist, it may well be necessary to hand the further investigation and interpretation over to either a surveyor, an architect or an engineer colleague, or to an outside consultant. The housing manager must, therefore, be aware of where trails can start and be able to identify when other skills are required to assist diagnosis.

Starting points

Occupied dwellings

With occupied dwellings, the most likely reason for instigating action is a complaint from the occupant. This may be by telephone, in person or in writing.

Voids

It may be that the housing manager is not dealing with an occupied dwelling but looking at a void property prior to allocation for letting. In these cases, there is little to base the inspection on and a potentially wider range of defects may be encountered.

Third parties

The information may come from a third party such as a neighbour. For example, it may be that a leaking gutter has not yet been noticed by a tenant, even though the consequent dampness is affecting the adjoining house.

Whichever of these is the starting point, the information that is then required is the same. The following list of things to consider may seem obvious, but it is so easy for any of us to omit an essential element that we see as peripheral until later in the process.

Is inspection necessary?

Finite resources mean that not all complaints can or should be inspected by housing managers. Every time a manager inspects, the money expended on that inspection is not being spent on a repair.

In some areas – e.g. in the London Borough of Croydon – tenants have been supplied with a manual that enables them to discuss and identify the defect with the assistance and prompting of the specially trained staff in the housing department. This empowerment of tenants not only reduces the need for the landlord to inspect before ordering repairs but gives the tenants direct involvement in the repairs process.

Increasingly, landlords are ordering repairs directly on complaint and relying on completion notes from the tenant to validate the contractor's claim for payment. This is a tempting area for future cost-cutting and, on the face of it, performance of a repairs service may not be harmed. After all, in assessing whether inspections are required, the landlord – particularly a large local authority or association – must assess the cost of failure. If this can be shown to be only marginally worse when fewer inspections are carried out, there may be little financial justification for carrying out such inspections.

The decision to inspect must, therefore, be based on the benefit of such an inspection. This, in turn, can only be based on the information available at the initial complaint stage. This information must then be interpreted by the housing manager on the basis of risk. This risk assessment takes into account the data and the potential for escalating costs as well as liabilities if the problem is not accurately diagnosed.

Recording information

Methodical recording of all events – even if this is in some code or in note form – is an essential discipline. It cannot be emphasised too often that many future disputes and difficulties with evidence arise out of a failure to take suitable notes. It is very easy to rely on an over-pressured memory or to assume that all will be solved swiftly and without dispute and that there will be no difference in recollection – in fact, these occur all too often. All the processes and actions of the housing manager must be tempered by this possibility of a potential dispute at a later date.

Computer databases are increasingly used and are extremely useful tools, reducing paperwork and providing better access with the potential for automatic filtering and initial diagnosis. However, these are tools as opposed to solutions, and the outputs are only as good as the inputs. Design and operation of any computer system needs sufficient checks and balances to ensure that the housing manager is provided with the relevant information.

Recording the initial information

Recording the complaint in a methodical way is essential. Cards and standard forms are used by many social housing landlords, recording the following information as a minimum:

- date and time that the information was received;
- the name and address of the informant, with telephone number if there is one;
- the exact words used by the informant; and
- any further information elicited by discussion of the problem.

This establishes the foundation from which the housing manager builds the response and, in particular – after considering all the background and recorded history available – should provide sufficient information for deciding whether an inspection is required or not.

Background information

Before setting off on an inspection, some basic information and equipment must be assembled. This starts with the record of the information referred to above but it will need to be expanded. Without background data, the visit will be less productive and possibly inconclusive, resulting in the need for a second visit. Repeat visits to look at the same problem should be avoided. The costs to the landlord are largely wasted, and the irritation to the complainant who sees only visits and no action is increased.

Address and identification code

This is needed not just so that you can find your way there but also so that records can be maintained and updated to enable a clear picture to be constructed and retained over the years. Each landlord has its own system of identification for recording the property history, usually on an electronic database. This may be entered or sorted by address, postcode, dwelling type, or tenant's name or rent roll reference. However this is done, recording of data and opinions in a methodical and consistent manner is essential.

Where a housing manager is familiar with a specific area, that local knowledge and experience should also be brought into consideration. This risks misdiagnosis of problems, however, especially if similar defects have apparently gone away in other properties; therefore, the local knowledge must be weighed with the other sources of information to give a balanced view.

Type of property

This should identify whether the property is a flat, a maisonette or a house. If it is a house, is it mid- or end-terrace; semi-detached or detached?

Size and location

A record should be kept of the number of habitable rooms and the number of storeys. If the dwelling is a flat, what is the size of the block and on which floor is it located? Is the block purpose-built or a conversion?

Building type

In the UK, there is a wide variety of building types (which are discussed in Chapter 1). Is the building known to be a traditional structure with brick walls, timber windows, a tile or slate roof on a timber sloping structure, and timber floors? If not, what sort of structure is it? This is seen on inspection, but the landlord's records of type allow some better pre-planning of the inspection. For example, if the building is of timber-framed construction then a defect such as a pipe leak that may, on first glance, seem minor could have a greater significance.

Occupants

Who is the tenant, how many people are there in the household and what are their ages? Are they in work?

If it is a large, overcrowded – and impoverished – household with occupants at home all day, this has a significant effect on the use of the building. The social factors cannot be separated from housing; disrepair is affected by them.

Tenancy agreement

What are the specific obligations of the landlord and tenant for this dwelling? It is assumed that for the majority of lettings managed by housing staff, the implied repairing obligations set out in section 11 of the Landlord and Tenant Act 1985 apply. This states the responsibilities of the landlord and the tenant. The landlord must maintain the condition of the structure and exterior of the property as well as ensuring that the installations – for example, for electricity, gas and water supplies – are in good order.

The landlord also has a duty of care under the Defective Premises Act 1972 to prevent injury to anyone using the property or damage to their belongings due to defects in the property. This includes defects that are known to the landlord as well as those that *should* have been known about. Where defects are caused by the tenant's failure to uphold any aspects of the tenancy agreement then the landlord is not liable.

Common parts

The existence of common parts should be a constant caution because of the different ways in which the law treats these areas even though, from a lay point of view, they may be indistinguishable in terms of the effect on both the building and the occupant.

After consideration of all the background information, the decision of whether to inspect or not can be made. There is always the danger of preconceptions based on the existing repairs history and, of course, the risk of repeating previous unsuccessful repairs based on inaccurate diagnoses.

There are no hard-and-fast rules which can be applied in all cases, but triggers based on the threats to the building or the consequences for the occupants should be utilised. Repeated reports of the same or similar defects should prompt questioning of why it has recurred.

Equipment

Having decided that an inspection is required, what equipment should be available for use?

Basics

The equipment listed here may not be used on every visit but represents the minimum which should be carried on every inspection in case it is needed.

Diary, clipboard, pen, pencil and *paper* may sound obvious, but how often do we trust our tasks to an imperfect memory and then fail to execute them fully?

Although tape recorders for dictation of notes on-site have clear advantages in terms of speed and volume of information, they also have some very severe disadvantages. It is likely that some of the notes required will include matters that should be kept confidential and which the occupant could find either puzzling or provocative. It is not always wise to enter debate with the tenant before any diagnosis has been thought through. The failure of a machine to record or the accidental erasure of the tape is another risk.

The perceived increase of efficiency and speed only becomes a reality where there are adequate support services available to transcribe the tape immediately on return to the office. The presence of background noise – from the occupants, the television or traffic – blurs the recording and makes accurate transcription difficult.

There is no permanent record of what was seen except for a typed transcript, which could contain minor typographical errors that significantly alter the record; e.g. the omission of the negative in describing the condition of a wall. The significance of contemporaneous notes is discussed later in Chapter 9.

Handheld computers are also available for site use. These can log the defect and the repair required; this is downloaded when back in the office, linking to the database and ordering systems. While these tasks are successful, all of the information collected on the inspection is rarely recorded; this includes the diagnosis that leads to the bald statement of what is wrong and what repair is required.

A good *torch* is an essential. Even with occupied properties, there are cupboards, dark corners and unlit areas where clues to defects could lurk. With void properties, it is likely that there is no electricity.

A small *screwdriver* or a *Swiss army knife* is required to probe into wood and for scraping material either to identify it or to take samples.

A *measuring tape*, at least 5m (16') in length, is needed to obtain overall measurements of rooms where this may be a consideration.

A small *sealable container* such as a tough plastic bag is needed for carrying samples such as insects or fungus for later identification.

Moisture meter

For any inspection where there is either a complaint of dampness or no specific defect identified prior to the visit – for example, on a relet – an electric moisture meter, however basic, is required. This measures the electric current passing through the material being tested between two probes. The greater the conductivity of the material, the higher the reading on the scale.

There is commonly assumed to be a direct relationship between the conductivity and moisture content of the material, but this is not always the case. The diagnosis of dampness is dealt with in Chapter 3, and it must always be remembered that a moisture meter is an aid to diagnosis and not the means of diagnosis itself.

Moisture meters commonly display readings according to Wood Moisture Equivalent (WME). Of all the materials found in buildings, wood is one for which a moisture content reading in 'per cent' is meaningful. It is generally accepted that wood rots when it is wetter than 20 per cent and that it is safe below this level. This is why moisture meters have a scale for wood. In any material other than wood, the meter gives readings of '% Wood Moisture Equivalent' (% WME). To put it another way: % WME is the moisture level in any building material other than wood expressed as a moisture content of wood. Therefore, a reading above 20 per cent in any building material indicates a hazardous condition that must be investigated further.

Some meters have a radio sensor that detects possible dampness below the surface (at a depth of about 19mm). This, again, can indicate a hazardous condition that must be investigated further.

Camera

Wherever possible, a photographic record of the inspection should be taken. This should supplement rather than replace the written notes. It is more important to have a photograph of the general condition and layout of the premises than to have a detailed close-up of a particular alleged defect (see also Chapter 9).

Use of drones

A recent development has been the experimental use of drones with cameras to photograph buildings from the air.

Chapter 5

The inspection

The purpose of the inspection is to obtain information which will be processed into a variety of reports and actions. It is an information-gathering exercise that is part of the overall process; it is not an end in itself. There may be limited opportunities to inspect for a variety of reasons; for example, the unwillingness of the occupier to give access, the inability of the occupier to be at home because of work or other commitments, etc. Repeat inspections to obtain information which should be obtained on a single visit are a waste of resources and a devaluation of the process.

This chapter focuses primarily on the vast majority of inspections where you will be met by ordinary people. However, there are also those cases where an occupant is known to be violent or where violence erupts during the inspection. Most landlords have their own records of tenants likely to be violent.

The experience of predecessors and more experienced colleagues is an extremely important resource for the newly arrived housing manager. Wherever possible, new managers should shadow or be inducted by colleagues whose greater experience can be passed on face-to-face with much greater effect than the written word.

Appointments and casual call systems

Many landlords with large stocks to maintain rely on a system involving staff calling on tenants without an appointment – and if there is no response, leaving a card inviting the occupant to contact them for another appointment. To the occupant, this may appear as a lack of respect, failing to take the status and the privacy of the family seriously and reducing their timetable and needs to a status lower than that of the landlord's employee. This does not suggest a cooperative climate in which to approach the visit and will almost without exception start a confrontational relationship which need not be there.

This system may in some cases appear to be an efficient use of resources, maximising staff time; but is it really? How many abortive calls are there for each successful call? What is the cost to the landlord of future confrontations aggravated by if not borne out of this casual approach to the tenants' needs?

Similar problems will result from a lack of punctuality. Unless the delay is unavoidable, tenants should not be kept waiting. After all, when the housing manager invites the tenant to meet at the office, the tenant is expected to arrive on time and to offer an apology if late.

There is no coherent or sustainable argument for bad time management and this, after all, is what the casual call system is. It is where the manager is unable to plan his or her workload and time to maximise productivity in the working day. Good time management is essential and this requires that inspections be made by appointment.

Diary systems

The working systems of landlords vary widely. A 'patch' system where each housing manager, or even a team, works on a particular area of the housing stock is recognised as a basically good system. However, with increasing use of short-term contracts and agency personnel, landlords may not be able to operate this system; or even if they do, frequent changes in personnel negate the advantages which should be derived from it. Where a person or team are dedicated to a patch, it will, of course, be more difficult to provide for substitution and cover for absences.

With these caveats, there are several mechanisms that can assist the office in achieving efficiency:

1. A *single centralised diary* for those carrying out visits/inspections. To be effective as a source for arranging appointments, such a diary needs to have time slots where people are available rather than the times when they have appointments. Changes in availability have to be logged centrally, consequently restricting the individual's flexibility in arranging further inspections as she or he has to refer back to the central control. Therefore, although this can appear to provide control and efficiency, in effect it reduces the freedom and flexibility of the individual and makes them less efficient.

2. *Individual diaries with no central liaison* are perhaps the most common diary system. The result is that only the operatives themselves can arrange appointments, leaving their colleagues and their customers frustrated and disenchanted. Additionally, there is no office record of where staff members have gone.

3. *Individual diaries linked and coordinated centrally* are a composite of systems 1 and 2. A central diary containing the individual diaries of all workers is updated daily, and there are time slots within which individuals have the liberty to arrange appointments as they arise during the day. By enabling individuals to arrange appointments within an

agreed space of time, individual control over workload is restored. An overview of the work and appointments of the whole staff is also maintained, enabling some swapping of tasks and visits in order to maximise resources. It requires dialogue between workers and a spirit of cooperation and team building. With remote access and smart phones, this can more readily be maintained.

Health and safety

It is an essential health and safety requirement that your office knows where you are and when you are due to return or to attend at another property. Guidelines on the safety of inspectors or others visiting premises away from their office have been prepared by the Suzy Lamplugh Trust. This trust was set up after the disappearance of estate agent Suzy Lamplugh while out of the office to show a prospective purchaser around a vacant house. Throughout the inspection, keep in mind your and your colleagues' health and safety. There is no room for heroics, and climbing on chairs or other unsatisfactory access equipment to get a better view is an unacceptable risk.

Making appointments

Preferably all appointments should be either made or confirmed in writing. Even where the appointment is arranged initially by telephone, the time and date should be confirmed. Where a precise time is not possible – for example, where a series of inspections of uncertain duration are anticipated – a time frame of not more than one hour should be given for the time of arrival.

How can this dictum work in practice for a range of inspections, from urgent to optional? In an emergency, written appointments are not appropriate. Here, the time should be arranged orally, preferably when the call reporting the problem is received. This is achievable where a centralised record of individual commitments is maintained (as in option 3 above).

For all other appointments, a letter or a telephone call followed by a confirming letter is essential. The letter generally suggests a time and date for the inspection but leaves unasked the questions of whether that time is acceptable and whether the tenant will be at home. This results in a vast waste of resources on the part of housing management staff. The use of prepaid reply cards to confirm or decline acceptance of the offered time is recommended. Non-receipt of a reply card confirming an appointment should always result in the appointment being cancelled and a further one offered.

Two problems arise in the use of such response-led appointments: namely the need for special consideration of tenants for whom English is not their first language and of those with sight or reading difficulties. Overcoming these problems and the potential cost of unused postage should not outweigh the benefits of prepaid cards. Tenants' special needs will still have to be catered for.

The documentation of the appointment is complete if this process is followed. This may seem of little relevance, but should litigation result, it could be of vital importance to demonstrate your response to a report by the tenant.

Conducting the inspection

The inspection will usually have been prompted by a complaint or report of a defect by the occupant. In most cases, this will potentially affect the interior of the dwelling, even if it is reported as an external defect only; for example, a slipped roof slate. There will be the exceptions of course, such as the broken garden fence or other external defect that has no potential to affect the interior or the structure of the dwelling. In these relatively rare cases, inspection of the interior may not be necessary; but in general, a comprehensive inspection is essential.

Introductions

The management of housing includes the establishment of a working relationship between the manager and the tenants. Managers visit tenants for a number or reasons, only one of which is disrepair. Nevertheless, it is worthwhile remembering that you are visiting someone's home. If it is for the first time, treat the visit as you would a first visit to a friend of your aunt's. You will usually be accepted and treated with as much respect as you show to the tenant. If you have visited previously, whatever the prior history, approach the home with respect and with candour.

It is often helpful to have a view of the neighbourhood and the adjoining houses before entering the particular property. There may be occasions when you are meeting with colleagues at the premises. The sight of a manager hovering outside is disconcerting, so as soon as you arrive, knock at the door to introduce yourself. Explain that you are having a brief look around the area or waiting for colleagues and that you will call back shortly.

Preliminaries

Before commencing the inspection proper, you should assess the character and type of adjoining premises, the general tone of the area and the age of the neighbourhood. For managers operating a patch or for those with current local knowledge, this will be familiar ground and will not need repeating. The environment in which the property is located has significance in determining the possible causes of disrepair, the likelihood of defects recurring, and the appropriate level and quality of repair required.

An initial assessment of the dwelling is useful. Is it of a standard type, likely to manifest similar defects to others in the neighbourhood of which you already have some knowledge? When was it first constructed and has it been substantially renovated or altered since? What is its likely lifespan before either demolition and redevelopment is required or major works of improvement and renovation have to be carried out? Again, managers operating a patch or those with current local knowledge should already have this information available.

The initial conversation with the tenant will inevitably contain much that is not directly relevant to the inspection; but issues such as late or inadequate benefit payments, high heating charges and recent family traumas or separations may all have an effect on the occupation of the dwelling or be a result of defects in the premises. Their relevance cannot be discounted but should be filtered.

Methodical procedure

The inspection must always be methodical. It can be difficult when an anxious occupier wants you to look at the kitchen, to insist on looking in the bedroom first. There needs to be a willingness to take an initial quick look at the problems presented by the tenant coupled with an assurance that you need to inspect the whole property in order to have an overview not only of the defect reported but also of its effect on occupation. This establishes a rapport and conveys to the tenant that you are listening to what they have to say.

The tenant's perception of the repair history can often provide an insight into previous actions that cannot be gleaned from the landlord's records. It can often give vital clues that previous attempts at repair or diagnosis have overlooked the real cause of the problem. The experience of tenants is all-important; after all, they live in the property.

Before you inspect, you must decide on how you will describe locations. Compass bearings can be used but these can lead to difficulties of interpretation by someone else, particularly if the directions appear on the repair order.

A simple basic rule that works well, no matter how you view a particular element during your visit, is to assume that you are facing the front of the property from the street, with rear, right and left consistently in their respective locations. It should also prove intelligible to repairers following after you. For example, from the rear garden looking towards the rear wall of the house, the elements on your left are described as though you are facing from the front – i.e. they are on the right. By rigidly adopting such a basis for locations, the relationship between interior and exterior and between various internal features can be more readily assessed in the subsequent report.

The order in which rooms and areas are inspected also needs to be consistent. Start on the top floor at the front right-hand room, move to the front left-hand room and then the rear left-hand room and rear right-hand room; i.e. in a clockwise direction.

Within each room, first look at the ceiling, then the floor, then the front, left, rear and right-hand walls, including the window(s), the doors and the fittings – such as fires, sockets, airbricks, etc. – within each of these elements.

Although this may seem an unnecessarily rigid method, it must be remembered that you will need to draft a report from your site notes, and that report may need to be interpreted and checked by others. If you jump from interior to exterior and introduce random elements into your notes, you will have to rearrange these and edit them before reporting or be faced at a later date with an embarrassing challenge to your methodology, or lack of it.

Using the senses

The use of a person's senses cannot be underestimated. Tools and equipment are merely there to assist you. Each inspection requires the application of all your senses: *smell* and *taste* for identifying damp, mould, sewage, gas leaks, etc.; *hearing* for water leaks, central heating faults, bird infestations, etc.; *sight* for discoloration, blemishes, missing components, etc.; and *touch* for dampness, looseness of materials, strength, operation of components, and so on.

Our senses will often give us more ready warning of a defect than instruments, which are then used to analyse, investigate and assist in the diagnosis of what has already been identified as a possible problem.

Identifying defects

The inspection, and consequently the notes taken, needs to be informed by diagnosis of the defect based on the information in Chapter 3 and Appendix II as well as other sources. There will be occasions when the housing manager cannot be certain of the cause of the defect or of the next stage of investigation required to get to the underlying cause. In such cases, the defect should be reported on the basis of the information the manager is certain of – probably the visible symptoms – together with any indicators as to underlying cause(s) that he or she feels may apply.

Inspection notes

The notes taken at the inspection will form the basis of future action and your response to the complaints of the tenant. The notes need to be sufficiently comprehensive to give you a picture of the dwelling and to record relevant actual, past and potential defects. They need to be written in such a way that you will be able to interpret them and use them as an *aide-mémoire* both on your immediate return to the office and at a later date, perhaps over a year later.

The notes should be written in ink, not pencil. They should be written legibly so that others can also read them. Abbreviations should be either commonly used or self-explanatory. Idiosyncratic acronyms should be avoided. The notes will include some descriptive matter to set the scene in which the defects themselves are to be viewed. A sample of notes from an inspection of a typical terraced house is set out in Table 5.1. Although these are typed for inclusion here, they are merely the notes taken on-site.

Numbering of rooms, areas and noted items is a good method for ordering the notes. This can help to shorten note taking where particular features are repeated and allows for ready cross-referencing; e.g. from interior to exterior.

In the sample notes set out in Table 5.1, a summary of the diagnosis is included. This diagnosis is not written out in full but, rather, indicates the questions that need to be asked and the routes of enquiry.

The inclusion of notes in square brackets can be a useful tool if the draft is to be immediately converted into a report. The format of reports is dealt with in Chapter 6, but it is important to remember that the end result should be kept in view throughout the process.

The notes taken at the time are the primary source of information from which future actions will flow. Accuracy of information collection is essential and time taken at this stage will pay rewards later.

Notes made during the inspection are contemporaneous notes and, as such, will form an integral part of evidence that may be presented in court at a later date. Giving of evidence is covered in detail in Chapter 9, but it should be remembered that reference to these notes may be the sole source for refreshing the memory in the witness box.

Table 5.1 Sample notes and diagnosis

Ref.	Note	Diagnosis
1	***First floor***	
1.1	*Front room (bedroom)*	
1.1.1	Mould growth to inside of steel [single-glazed] window frames	Steel windows are very vulnerable to condensation and single glazing means that more moisture will condense on and run down from the glass on to the frame and sill. Water condensing out on the frames will give rise to mould growth. Is there any other source of moisture? |//
		(Continued)

Table 5.1 (Continued)

Ref.	Note	Diagnosis
1.1.2	Mould growth to ceiling to front wall junction	The location of the mould at the wall junction suggests condensation due to cold bridging. Is there any other source of moisture? If so, the area would be expected to be damp.
1.1.3	Outlets from drain channels to middle rail of window filled	These channels should drain the water condensing on the inside of the windows.
1.1.4	Timber subframe to window rotten and opening at joints	The rot is caused by moisture which, in the early stages, will cause joints to swell and open up. The inside of the window is likely to get this moisture from the condensation running down the inside of the window glass and frame, but is there any rot on the outside of the frame? Is the inside more rotten (soft) than the outside?
1.2	*Rear left-hand room (bedroom)*	
1.2.1	Mould growth to inside of steel [single-glazed] window frames	Repeat of 1.1.1.
1.2.2	Mould growth to ceiling to front wall junction [Calor gas heater in front cupboard – no gas cylinder]	The gas heater, if used, will discharge considerable water vapour into the dwelling. The discharge will affect the whole dwelling, not just the room in which it is used. Has it been in recent use?
1.2.3	Outlets from drain channels to middle rail of window filled	Repeat of 1.1.3.
1.2.4	Timber subframe to window rotten and opening at joints	Repeat of 1.1.4.
1.3	*Rear right-hand room (bathroom/WC)*	
1.3.1	Glass to window cracked [possibly hit by stone from outside]	Cracks in glazing with an impact point may be due to damage not falling within the obligation to repair. Can the cause of the cracked glass be seen?
1.3.2	Mould growth to steel [single-glazed] window frame	Repeat of 1.1.1.
1.3.3	Outlets from drain channels to middle rail of window filled	Repeat of 1.1.3.
1.3.4	Timber subframe to window rotten and holed externally	Repeat of 1.1.4, but in this case there is damage externally, suggesting a cause additional to condensation.

Ref.	Note	Diagnosis
1.3.5	Mould growth to ceiling to right-hand 1m and extending along rear wall junction to left-hand corner [air grille rear wall, high level, papered over]	Repeat of 1.1.2. The blocking of the air grille will reduce the discharge of moisture-laden air from the room, increasing the likelihood of condensation. The opening of the grille will probably reduce the temperature in the room and will not be as effective as a powered extract fan.
1.3.6	WC pan loose on floor [no leaks]	Why is the pan loose? Is the floor rotten? Is the instability causing leakage that could cause rot or other damage, or even allow soil to discharge on to the floor?
1.3.7	Washbasin loose on left-hand wall and basin cracked [small indentation looks like impact damage, recent]	Is the crack clean and fresh? Does it have any indications of impact? These may affect whether it is the landlord's liability to repair.
1.3.8	Door holed at base and centre adjoining lock stile [probable deliberate damage – trying to force open]	When was the damage caused and how? This may affect whether it is the landlord's liability to repair.
2	**Ground floor**	
2.1	*Front left-hand room (living room)*	
2.1.1	Dampness (75 WME) to front wall and plaster perished and loose to average height of 850 mm above floor level, with section of plaster missing (approx 300 × 200) to left of window [no injection holes outside, solid brick wall, no visible damp-proof course, sharp reduction in meter readings at top of damp area]	What is the moisture meter reading? Is there a sharp change in readings? Are there any visible signs of deterioration to decorations or plaster? Is the skirting board of timber and is it damp and/or rotten? All these help to assess the possible source and the seriousness of the defect.

What is the wall type? Is there a visible damp-proof course? Is the plaster hard or soft? These help to indicate possible areas of failure of the structure and the direction of any further investigations.

The information obtained indicates a problem of dampness due to an ineffective barrier to rising damp. |
| 2.1.2 | Dampness (20 WME) to rear internal wall to height 1 metre [solid block wall, no visible plaster damage, no visible damp-proof course, gradual reduction in meter readings at top of damp area, no mould but tenant claims that the skirting gets mouldy] | Repeat of 2.1.1. In this case, the information indicates condensation as the most likely cause, due to lower temperature of the wall where it is in close proximity to the ground. |

(Continued)

Table 5.1 (Continued)

Ref.	Note	Diagnosis
2.1.3	Mould growth to inside of steel window frames [no mould to reveals or to plasterwork and decorations adjoining]	Repeat of 1.1.1.
2.1.4	No permanent ventilation to room – gas fire fitted to fireplace	The Gas Safety Regulations require landlords to ensure that all gas appliances are safe for use. In general, all appliances other than some balanced flue units require permanent ventilation. Annual checks are mandatory.
2.1.5	Door binds on frame and hinges loose	How did the door become loose? Is there evidence of deliberate damage? Is the frame defective or have the screw holes become oversized and worn through old age?
2.1.6	Outlets from drain channels to middle rail of windows filled	Repeat of 1.1.3.
2.1.7	Timber subframe to front window rotten and opening at joints	Repeat of 1.1.4.
2.1.8	Timber subframe to left-hand window rotten and holed	Repeat of 1.1.4.
2.2	*Rear left-hand room (kitchen)*	
2.2.1	Peeling paintwork to ceiling and mould growth to rear left-hand corner	Repeat of 1.1.2. The non-operation of the fan will reduce the discharge of moisture-laden air from the room, increasing the likelihood of condensation.
2.2.2	Mould growth to PVC window frame [double-glazed]	PVC windows with double glazing are not usually vulnerable to condensation. However, the non-operation of the fan will reduce the discharge of moisture-laden air from the room, increasing the likelihood of condensation.
2.2.3	Electric extract fan not operating [electric outlet indicator light on]	Why has the fan failed? Has it been disconnected by the occupants? Is the electricity on – i.e. has any prepayment meter been charged?
2.2.4	Loose floorboarding to front of centre adjoining doorway	This is a ground floor and the subfloor timbers are always vulnerable to dampness either from the ground or from the dwelling. Ventilation of the void is essential. Are the boards loose because the supporting timbers are rotten? This may require further investigation which cannot be carried out at the initial visit.

Ref.	Note	Diagnosis
2.2.5	Hinge stile of door broken away at mid-height	When was the damage caused and how? This may affect whether it is the landlord's liability to repair. It may have been an attempted break-in.
2.2.6	Outlets from drain channels to sill of window filled	Repeat of 1.1.3.
2.2.7	Waste trap to sink leaking in cupboard and shelving to cupboard saturated and disintegrating	If this is left unrepaired, further rapid deterioration of the cupboard will occur.
2.3	**Hall**	
2.3.1	Damp-staining to soil pipe duct within rear cupboard indicates leakage from pipework serving upper unit	Without access to the upper unit, it is not possible to assess the cause. It may be a defect caused by misuse in the flat above. Further investigation is required.
3	**Exterior**	
3.1	*Front elevation*	
3.1.1	No projecting drip to head of window subframes	This may relate to the window and ceiling/wall defects noted in each room. If there is no projecting drip, water will tend to run into the junction between the head of the window and the wall. This could cause dampness internally and by making the wall damp, increase the cold bridge effect, in turn increasing the likelihood of condensation.
3.2	*Left-hand flank elevation*	
3.2.1	Metal drip to head of window directs water on to timber subframe	This may relate to the window defects noted in each room. The direction of water on to the external sill may be the primary cause of the rot to the sill.
3.3	*Rear elevation*	
3.3.1	Metal drip to head of windows directs water on to timber subframe	Repeat of 3.2.1.
3.3.2	Damp-staining to brickwork on line of soil pipe further indicates leakage (see 2.3.1 above)	This reinforces the indication that there is a long-standing leak from the pipe and that further investigation is required. See 1.5.1.

Box 5.1 Example

An inspection of sample property, resulting in the notes set out in Table 5.1, might proceed as follows.

The manager arrives, with or without colleagues, and introduces her/himself to the tenant. The manager explains the purpose of the inspection and that in order to put the matters in context, some appreciation of the area and adjoining buildings is required. The manager briefly looks around the immediate neighbourhood and at the relationship of the subject property to the adjoining buildings or sites, noting the addresses of those adjoining sites or buildings, or their location where no address exists.

Returning to the premises, the manager again knocks and explains that an inspection of the interior will be followed with inspection of the exterior and possibly some discussion with the tenant. The tenant should be invited to confirm the nature of the reported defect but the manager should resist chasing after a series of randomly located defects rather than inspecting the dwelling on a methodical basis.

Armed with (a) the originating information and (b) any elaboration of this or additional matters raised by the tenant, the manager starts to inspect.

Following my preferred order of inspection, the inspection starts in the top front room. The inspection of this room follows the pattern for all interior spaces and, therefore, the following description of this process has a general application.

The first tools to be used are the eyes. Locate the room mentally in the dwelling and note whether there are any other rooms on the same floor and how they compare in overall size to the subject room. Look around the room and take in its general arrangement and how the elements fit together to check that the preconceived format (this is held in your head, not on paper) will be suitable. It may, for example, be necessary to add a further category to allow for a bulkhead or offset wall, or windows on more than one elevation.

It may be helpful to sketch the layout. If so, a few simple rules need to be applied: wherever possible, use squared paper as it makes sketching much easier; sketch the external outline of the dwelling in plan with the front external wall at the top of the page, leaving a generous margin for any notes or features; then infill with the rooms. Draw the rooms so that you will be able to plot, and later recognise, the shape and features of the space.

Mentally compile a list of the elements and start with the ceiling. Observe the layout, assess its possible construction and its condition.

Look for 'trigger defects' (see Appendix II) and record them, including their approximate location. Always look into built-in cupboards. Because they are often built to enclose awkward junctions between walls and ceilings or in corners of rooms, they can conceal serious defects not yet reported by the tenant but which will worsen if ignored.

The description should be brief but contain the salient points. This can best be illustrated by the example in Table 5.2, which is not an exhaustive list of possible elements of construction but does give an indication of the information that can be gleaned.

Table 5.2 Ceiling

Ceiling	
Item	Possible description
Decoration	Painted on finish surface
	Painted on paper
	Papered
	Undecorated
Finish	Plasterboard
	Plaster (lath and plaster)
	Other boarding
Structure	Ceiling joists
(assessed, not necessarily seen)	Floor joists
	Concrete roof
	Concrete floor

Follow the ceiling with the floor. Note how it is covered, whether it is liftable and, if so, what the structural covering is and which way this indicates the structure of the building.

Next, observe the walls, including the elements such as windows and doors contained within them. Open windows and doors to check their operation and adequacy of fit. Lift doors gently by the handles to check for movement on hinges. Assess the effectiveness of external doors and windows to minimise draughts and water penetration. Use views from the windows to observe low-level roofs and pipes. Test the condition of the window sill and sub-sill. Although part of the exterior, it is convenient to include these elements with the interior. Refer to the checklist in Appendix II for additional information to be gathered based on particular visible defects.

Proceed with the other elements to complete the picture of the room. Before leaving the room, take a look around to refresh your overall impression. This is the opportunity to take photographs to act as *aide-mémoires* (see Chapter 4).

Continue to other areas/rooms in the house following a methodical order and repeat the process. This should not take long in rooms/areas where there are no defects. The descriptions can be brief and include acronyms and abbreviations. With a relatively small amount of self-training and practice the whole process can be refined and streamlined, and to inspect one room will take significantly less time than it takes to read about the process in this book!

Proceed floor by floor to the lowest floor. Include, but identify as distinct from the demised premises, the common parts providing either cover, access to or escape from the premises. Areas to which access is necessary to check service meters or to turn off supplies in an emergency should be included.

From the lowest floor interior, where you should finish in the rear room, go to the rear exterior. Deal first with the rear wall(s) of the dwelling and the building within which it is contained. Log each wall of the elevations. Often there will be separate parts of the rear, usually containing the kitchen and bathroom. These need to be identified separately to ease cross-referencing to the interior and to ensure that all elements of the building are included. The side (flank) elevations are then dealt with, followed by the front. Again, differentiate between the main walls and subsidiary structures such as bays, porches and any lean-to.

The inspection can then move to the site, the front and rear gardens and any access paths. The extent to which these are included in the landlord's obligation to repair is discussed elsewhere (see Chapter 7). Notwithstanding the obligations on the parties, the description and condition of paths, fences, gates and access areas as well as outhouses, garages, etc. should be recorded.

Cross-refer interior to exterior before leaving the premises to ensure that you have followed the trail as far as possible and that you have linked internal defects to external disrepair wherever possible.

There will always be defects that have no immediately obvious cause. These do not require resolution at the premises but may be matters that require either access equipment or specialist advice. Try to assemble as much information on these items as possible so that future action can be planned and discussions back in the office made as effective as possible.

Before finally departing, check through your notes and ensure that all areas have been inspected or note those areas where no inspection has been carried out. Make sure you, at least, can read what you have written.

Have a final discussion with the occupant giving, wherever possible, an indication of the time it will take to analyse your findings and decide on what action, if any, to take. Try not to promise works or other action unless you are 100 per cent confident that you will be able to deliver, and only give timescales that are certain to be achieved.

Timescales

To give dates or undertakings which may not be delivered may bind the landlord and give ammunition for any opponents should the matter proceed to litigation.

There is a widely held belief among tenants, particularly in the public sector, that any timescales will be exceeded and that any undertakings will be broken. This is not always a justified view but, nevertheless, the perception is thus. Accordingly, although occupants may consider that the works they perceive as essential should be carried out immediately, they will in my experience be more concerned that what they are told is going to happen will in fact happen. While not wishing to suggest evasive responses, it is probably better to offer a response in a few days after you have considered the action required.

Timescales within which local authorities make repairs are required to be published (section 167 of the Local Government and Housing Act 1989). Even without these timescales, many social landlords set their own guidelines and targets, which are published (see Chapter 7). Although adherence to these is important, they can only apply if the repairs are not complex or the causes uncertain.

Do-it-yourself

Sometimes it is very tempting to carry out a minor repair oneself. For example, where an extract fan is not working, a simple test is to change the fuse. If this allows the fan to operate then there is no point in removing the good fuse to replace the defective one, just to maintain the defect. However, there still remains the question of why the fuse blew. The defect is not fully repaired until the cause of the failure has been identified and/or the problem repaired with confidence of its future safe performance. Any such repairs must be noted so that (a) there is a record of your action and (b) others who will follow you will know what has been tried. There may be matters – such as electrics in a kitchen or bathroom, or gas installations – where only suitably qualified persons should undertake works.

Avoid the temptation to get things going or to carry out repairs which could fail. First, a housing manager is not a trained building operative and will be exceeding her/his job description by carrying out repairs. Second, if the repair is one for which the landlord, who may be the employer, is not liable, the execution of work by the manager may give rise to a claim that the additional liability has been accepted. Third, should the repair fail, you and your employer could be liable for that failure and any consequential losses.

Finally, a full assessment of the risks and the liabilities must be carried out. This will include legal liabilities (see Chapter 7) and require fuller specification of the remedial works in due course.

Part 3

Post-inspection practice

Chapter 6

Reporting

The reporting process, like note taking on-site, is an essential ingredient to effectively deal with disrepair and to minimise the risk of later recurrence of defects or further complaints.

It is not the function of this book to set out 'master procedures' for housing managers. To do so would deny all the experiences of individuals and the diverse but often equally effective processes already in place. The benefits of following controllable and comprehensive procedures can only really be illustrated with examples. Therefore, this chapter deals with aims and objectives of reporting using examples rather than setting these out as the only methods which are acceptable.

Customers

Reporting is a means of communicating what was seen on-site and informing the future actions or considerations of others and of the report's author. Therefore, before starting to write the report, the manager will need to establish who are his or her readers or customers. The TQM (Total Quality Management) technique uses the concept of customers. A simple question and answer helps to explain and define this concept: Q. Who are your customers? A. Anyone who receives a product or service from you.

These customers will be both internal and external. The internal customers are the manager's colleagues who have to take action based on his or her information. This may, of course, include the manager's future actions and responses, and those of other allied departments of the organisation such as Legal, Building Works, Technical Services, etc. External customers will probably be limited to the end user (the tenant) but may also include insurers and management committees who will have demands of their own which need to be satisfied by the adopted process.

The findings

By rechecking the initial note of complaint, it is possible to modify the perception of what was initially thought to be clear in the light of the information obtained from the visit. For example, the initial report may have referred to a specific defect which, on inspection, was found to be one of a number that are of equal or superior concern. The tenants may have indicated the involvement of their own advisers and possibly an expert, or suggested that litigation was being contemplated where no previous indication was given.

The findings of the inspection may also send signals that litigation is a possibility. There may be defects which, although serious, the housing manager knows are not part of the landlord's current scheme for repairs and will, therefore, take a revision of policy to resolve. Such matters as condensation-caused dampness can easily give rise to litigation long before any changes in policy, let alone funds for works becoming available.

It may not be only the possibility of litigation which sets the format and content of the report. The defects found may be ones which the housing manager knows will be addressed in a future rehabilitation scheme or a larger package of improvements. Such major schemes will often preclude expenditure on interim repairs which will be undone by the major scheme. Nevertheless, there may be a genuine need to address the repair of defects in the short term. This is especially the case with estate-wide improvement schemes which depend on the vagaries of the public sector housing finance system. This can often make the timing of such schemes seem more a matter of chance – rather like a lottery – than planned.

In such cases, it may be that the housing manager has to campaign within the landlord's organisation for expenditure in advance of the major scheme.

Format

In order to respond to this variety of customers, the format of the report must be as flexible as possible. To achieve this, reports are best produced as a series of components that can be arranged and rearranged to suit the changing needs of the customers.

The core section of the report contains the conditions found on-site. This will, however, need to be divided into those matters that require action and those which are for the record and may need to be drawn on at a later date. There may also need to be an overview of the conditions found and preliminary views on the liability of the parties as well as recommendations for future action and time priorities.

The use of standard formats for each of these components enables a higher level of quality control and maximises use of the scarcest resource, the housing manager's time. The following component formats are suggestions, and as stated above, their inclusion here is to illustrate their content rather than to set out rigid rules that have to be applied.

Core section

The report's core section summarises the information gathered on defects and identifies the repairs that are required. There will be occasions when the housing manager is not certain either of the cause of the defect or of the remedial works required. In such cases, the defect should be reported on the basis of the information of which the manager is certain – probably the visible symptoms – together with any indicators as to the underlying cause in which they feel confident. The repairs required may be limited at this stage to further investigation, possibly by an expert (see Chapter 9).

Using the example of the notes set out in Chapter 5 (see Table 5.1) along with the descriptions in Chapter 2 and Appendix II, it is possible to provide a schedule of defects and a brief outline of the repairs that may be required. A sample schedule is set out in Table 6.1. Some of these items – for example, 2.1.1 – require an expert to diagnose the cause of the defect and specify the repair.

Table 6.1 Schedule of defects and repairs

Note: all directions taken facing the front of the premises from the street

Ref.	Defect	Repair
1	**First floor**	
1.1	Front room (bedroom)	
1.1.1	Mould growth to inside of steel window frames	Following repairs to window subframes, remove mould and make good to decorations.
1.1.2	Mould growth to ceiling to front wall junction	Remove mould and make good to decorations. (Note: This is a shorthand for a more detailed specification depending on the exact circumstances.)
1.1.3	Outlets from drain channels to middle rail of window filled	Clean out drain outlets and make good to decorations.
1.1.4	Timber subframe to window rotten and opening at joints	Renew subframe completely and make good to plaster and decorations disturbed.
1.2	Rear left-hand room (bedroom)	
1.2.1	Mould growth to inside of steel window frames	Following repairs to window subframes, remove mould and make good to decorations.
1.2.2	Mould growth to ceiling to front wall junction	Remove mould and make good to decorations. (Note: This is a shorthand for a more detailed specification depending on the exact circumstances.)

(Continued)

74 Post-inspection practice

Table 6.1 (Continued)

Ref.	Defect	Repair
1.2.3	Outlets from drain channels to middle rail of window filled	Clean out drain outlets and make good to decorations.
1.2.4	Timber subframe to window rotten and opening at joints	Renew subframe completely and make good to plaster and decorations disturbed.
1.3	*Rear right-hand room (bathroom/WC)*	
1.3.1	Glass to window cracked	Renew glazing, make good to putties and decorations.
1.3.2	Mould growth to steel window frame	Following repairs to window subframes, remove mould and make good to decorations.
1.3.3	Outlets from drain channels to middle rail of window filled	Clean out drain outlets and make good to decorations.
1.3.4	Timber subframe to window rotten and holed externally	Renew subframe completely and make good to plaster and decorations disturbed.
1.3.5	Mould growth to ceiling to right-hand 1m and extending along rear wall junction to left-hand corner	Remove mould and make good to decorations. (Note: This is a shorthand for a more detailed specification depending on the exact circumstances.)
1.3.6	WC pan loose on floor	Securely fix pan to floor.
1.3.7	Washbasin loose on left-hand wall and basin cracked	Renew basin and securely fix to wall; run mastic seal to wall tiling junction.
1.3.8	Door holed at base and centre adjoining lock stile	Reface door and make good to decorations.
2	**Ground floor**	
2.1	*Front left-hand room (living room)*	
2.1.1	Dampness (75 WME) to rear wall and plaster perished and loose to average height 850, with section missing (approx 300 × 200) to left of window	Remove skirting board and check condition/existence of damp-proof course. Assess and execute necessary remedial work including renewal of plaster, skirtings, etc.
2.1.2	Dampness (20 WME) to front wall to height 100	Remove skirting board and verify that dampness is limited to condensation.
2.1.3	Mould growth to inside of steel window frames	Following repairs to window subframes, remove mould and make good to decorations.
2.1.4	No permanent ventilation to room – gas fire fitted to fireplace	Carry out checks in accordance with Gas Safety Regulations and install permanent ventilation if required. Make good to plaster and decorations disturbed.
2.1.5	Door binds on frame and hinges loose	Adjust and rehang door; make good to decorations.
2.1.6	Outlets from drain channels to middle rail of windows filled	Clean out drain outlets and make good to decorations.

Reporting 75

Ref.	Defect	Repair
2.1.7	Timber subframe to front window rotten and opening at joints	Renew subframe complete and make good to plaster and decorations disturbed.
2.1.8	Timber subframe to left-hand window rotten and holed	Renew subframe complete and make good to plaster and decorations disturbed.
2.2	*Rear left-hand room (kitchen)*	
2.2.1	Peeling paintwork to ceiling and mould growth to rear left-hand corner	Remove mould and make good to decorations. (Note: This is a shorthand for a more detailed specification depending on the exact circumstances.)
2.2.2	Mould growth to PVC window frame	Remove mould and make good to decorations. (Note: This is a shorthand for a more detailed specification depending on the exact circumstances.)
2.2.3	Electric extract fan not operating	Test, overhaul and repair fan and leave in full working order.
2.2.4	Loose floorboarding to front of centre adjoining doorway	Lift boarding and inspect; repair as necessary; securely fix down; make good to floor finish.
2.2.5	Hinge stile of door broken away at mid-height	Replace door and make good to decorations.
2.2.6	Outlets from drain channels to sill of window filled	Clean out drain outlets and make good to decorations.
2.2.7	Waste trap to sink leaking in cupboard, and shelving to cupboard saturated and disintegrating	Remake joints to waste outlet, renew shelving complete and make good to any decorations disturbed.
2.3	*Hall*	
2.3.1	Damp-staining to soil pipe duct within rear cupboard indicates leakage from pipework serving upper unit	Inspect upper unit. Check and test pipework and repair as necessary; make good to boxing and decorations.
3	**Exterior**	
3.1	*Front elevation*	
3.1.1	No projecting drip to head of window subframes	Following renewal of frames, fit projecting drip.
3.2		*Left-hand flank elevation*
3.2.1	Metal drip to head of window directs water on to timber subframe	Following renewal of frames, fit projecting drip.
3.3	*Rear elevation*	
3.3.1	Metal drip to head of windows directs water on to timber subframe	Following renewal of frames, fit projecting drip.
3.3.2	Damp-staining to brickwork on line of soil pipe further indicates leakage (see 2.3.1)	See 2.3.1.

Supplementary sections

These can vary considerably and require tailoring to the circumstances of each case, but it is possible to establish a menu from which the housing manager can choose the appropriate formula to supplement the core section, dealing with particular aspects such as alleged tenant misuse, previous incomplete or unsuccessful repairs, and earlier incorrect diagnoses.

Record note

This is required for file information and to put both the site notes and the report in context for other staff. It includes a general description of the premises and notes on the overall conditions found. It should include reference to photographs taken and any sketches made that serve to illustrate either the general setting of the property or particular defects.

A record note should be produced in a typed form rather than being handwritten. It should be brief and seek to answer a series of questions as set out in Table 6.2. It does not need to be in a questionnaire format but can be produced in normal prose. Like a jigsaw, the relevance of each piece of information is only fully appreciated when the whole is completed. Although each aspect may appear simple and not contentious, it is shown in Chapters 7 and 9 that all these aspects have relevance.

Table 6.2 Sample record note

Question	Example
Address	16 View Road, VW7 8AS
Who inspected	P. Reddin
Date of inspection	23 June 1995
Approx. time of inspection	15.30
Weather conditions	Heavy rain
Type of dwelling	Ground-floor flat, 2 beds, living room, kitchen and bathroom
Type of block	4-storey purpose-built
Location	Mid-terrace
Approximate date of construction	c.1890
Approximate date of substantial modernisation works	c.1975
Anticipated life	30 years
Number of occupants and approx. ages	Adults 1M, 2F
	Children 2–10 1M
	Children 0–2 1F
Heating	Gas fire in living room
Ventilation	Airbricks to kitchen and bathroom

Landlord's liability

In order to assess the future action that needs to be taken, the liability of the landlord must be addressed. The primary source of information about the liabilities of the landlord will be the tenancy agreement. This ought to be read prior to the inspection but will need to be referred to again in the light of any defects found. Reference also needs to be made to the statutory provisions and common law (see also Chapter 7).

The obligations of the landlord need to be related to the defects set out in the core section of the report. Not all the defects will fall within the obligation to repair and these must be identified. It may be that some items of disrepair will be rectified by the landlord as a matter of policy, but such works must be undertaken with authority; otherwise, doing this may set a precedent that commits the landlord to future liability.

An indication of the urgency of the repair should also be included. This is discussed in Chapter 7 in relation to statutory provisions under the 'Tenant's right to repair' and also in relation to guidance and good practice.

Tenant's liability

The tenancy agreement will usually set out the obligations of the tenant. In public sector social housing, this is usually restricted to the obligations not to commit waste, to use and keep the premises in tenantable repair and not to commit nuisance. There may be other obligations including a prohibition on criminal use, on interference with quiet enjoyment of neighbours, etc. These are matters that a housing manager may wish to include in a report, but they are beyond the scope of this book.

Box 6.1 Repair obligations: plasterwork

One of the more contentious areas over the past few years has been liability for the repair of plaster. The Landlord and Tenant Act 1985 (section 11) puts an obligation on landlords to 'keep in repair' the structure of a property, and the lack of consensus centres around whether plaster is part of the structure.

My own view has always been that repair of plaster should be the liability of the landlord. Although not structural itself, plasterwork forms a whole (e.g. a wall) together with other elements that are structural, and it is necessary so that a tenant can apply a decorative finish. In some circumstances (such as when dealing with dampness), it is essential to renew the plaster as part of the remedial works to achieve a proper repair. It also provides a small amount of thermal insulation to the internal surface of the exterior of the building.

> In the courts, however, various decisions have been given on the nature of plaster in relation to a building's structure. Some cases concluded that plaster is part of the structure and therefore the landlord is liable for repair (e.g. *Quick* v *Taff-Ely Borough Council* [1986]; *Hussein* v *Mehlman* [1992]; *Staves & Staves* v *Leeds CC* [1991]). In another decision, 'structure' was viewed as 'those essential elements of the dwelling house which are material to its overall construction' while internal wall plaster was 'in the nature of a decorative finish' so not structural (Mr Recorder Thayne Forbes in *Irvine* v *Moran* [1992]). This definition of 'structure' was approved in *Marlborough Park Services Ltd* v *Rowe* [2006], but the point on plaster was not. This changed again in the case of *Grand* v *Gill* [2011], the decision being that plaster is a constructional finish on which decoration is applied. It is, therefore, to be considered part of the structure with liability for repair falling with the landlord.
>
> Of course, another aspect relates to the cause of damp and whether the landlord has any obligation to remedy damage to plaster caused by condensation due to the actions of the tenant. It is useful at this stage to be reminded of a 2014 County Court case (*DR* v *Southwark LBC*): a claim for disrepair involving damp, saturated plaster to the external bathroom wall. The cause of disrepair was thought to be a mix of damp penetration and condensation damp. Investigation showed no water penetration to the core of the wall; the tenant agreed that condensation was a major cause of the problem but argued that it was the landlord's responsibility to repair plasterwork. The outcome was that the landlord received a fine.

Reference to further action

The housing manager can draw preliminary conclusions from the information obtained on-site. Based on this information, required future action can be assessed and a plan of action implemented. The conclusion needs to include at least an initial list of other departments or persons who should be aware of the inspection and report. It does not necessarily mean that they need to see the report at this stage.

Chapter 7

Priorities

The urgency for future action will have been determined initially by the housing manager as part of the report (see Chapter 6). The action could be the execution of repairs or it could be further investigation or guidance. As seen in Chapter 5, there will be many occasions where accurate diagnosis of the cause of a defect, and consequently the specification of the repair required, will either be beyond the competence of the housing manager or not possible at the time of initial inspection. Housing managers are not engineers, surveyors or architects and are unlikely to possess sufficient technical skills to diagnose all defects fully. An inaccurate diagnosis could increase the liability of the landlord and confuse issues. It is essential that housing managers take a realistic and modest view of their technical skills and that there is sufficient technical backup available to them where this is required.

External advice

The particular skills required to diagnose the defect may be available within the housing manager's organisation, but if not then external assistance and advice may have to be sought. In deciding the agency and person to whom a referral will be made, whether this is internal or external to the organisation, consideration must be given to the following factors.

If the further investigation is purely of a technical nature, then referral should be on the basis of the person's or agency's track record of dealing with such matters. Often, the organisation will maintain a list of consultants to whom particular categories of problems can be referred.

Where the further investigation will also require an expert opinion, which could be open to challenge in any subsequent court hearing, the selection is more critical. The referral will not be to an agency but to a specific individual (see Chapter 9).

Reference can also be made to the 2002 publication by the Department for Transport, Local Government and the Regions entitled *Housing Disrepair Legal Obligations: Good Practice Guidance*.[1]

Specification for repairs

Only after carrying out the further investigations and receiving the advice sought, either from internal colleagues or from external sources, can the final list of repairs be drawn up. There is no set format for this and it depends on the structure of the organisation how the works will be implemented. The repair will, however, now be sufficiently well specified to allow works to be ordered. The orders should be unambiguous and readily understandable.

Assessing priorities

The priority for repair needs to be assessed both on an individual and on a strategic basis. There ought to be a stated relationship between planned and reactive maintenance and an awareness of possible forthcoming major schemes of renovation. Dissemination of strategic decisions and possible schemes within the landlord's organisation is required. The existence of plans for major refurbishment, together with their current status and anticipated timescale, should be known to all housing managers.

General factors

Having assessed the liability for repair (see Chapter 6), the priority for the particular repair should be established. Priority is determined by the following factors (discussed below):

- the statutory obligations of the landlord;
- the guidance available in codes of good practice;
- the planned works of repair already set in motion;
- the anticipated time before major works of renovation are executed or the property redeveloped;
- value for money.

Traditionally, the availability of resources will also influence the priorities. The courts have generally not accepted the plea of 'poverty' to excuse a landlord from obligations to repair. However, in the case of major works which could have far-reaching effects on finance, the courts have been reluctant to include these in the landlord's repairing obligation.

As an example, the problem of condensation-caused dampness in local authority housing is one which seriously affects not just the enjoyment of those dwellings but also the health of the occupants. In civil actions, the courts have generally not accepted that the works required to remedy this defect (a combination of heating, insulation and ventilation) are within the landlord's repairing obligations as contained in section 11 of the Landlord & Tenant Act 1985, notwithstanding the actual and potential damaging effects on the occupants and the structure of the landlord's property. Nevertheless, by using the Environmental Protection Act 1990, tenants have been able to secure orders from the magistrates' court requiring works to remedy condensation dampness and mould growth.

The inclusion of an improvement within a repairing obligation will be allowed only where the marginal cost is minimal. In this way, the availability of finance can have an influence on the liability of the landlord to execute a particular repair.

The purpose of establishing priorities is to maximise the limited resources of the landlord to deal most effectively and equitably with the often conflicting demands of the occupants of the dwellings. In most cases, occupants would like all the repairs executed immediately, but equally they will accept that this may not be possible.

The factors listed above require analysis in greater detail to understand their influence and relative importance in the establishment of the priority. To a large extent, these priorities are set by the organisation at a strategic level. The housing manager needs to be aware of the criteria for these priorities as well as using them to determine the extent of work and response time to a particular case. These factors are now considered in turn.

Statutory obligations

Section 11(3) of the Landlord & Tenant Act 1985 limits the repairing obligation of the landlord by taking into account the age, character, location and prospective life of the dwelling. To give two extreme examples: the court will not order renewal of a roof covering where the whole building is to be redeveloped within a year; conversely, the court will have little hesitation in ordering the renewal of a roof on a building in a neighbourhood of high-quality dwellings with a prospective life of over 60 years. Unfortunately, extreme examples are rare. It is the vast grey area in the middle which requires examination to attempt to establish some principles.

Local authorities have powers to take action if there are any risks to the health and safety of occupiers or visitors to dwellings under the Housing Act 2004. Up until 2004, the fitness of dwellings was assessed using the Housing Fitness Standard (section 604 of the Housing Act 1985). This set out the basic requirements for dwellings to be considered acceptable for occupancy. A main criticism of the Fitness Standard was that it did not distinguish between dwellings that were defective and dwellings that had genuine health and safety risks for the occupants. Following a process of review and consultation, a new system was introduced in the Housing Act 2004 and brought into effect by the Housing Health and Safety Rating System (England) Regulations 2005. This system (the HHSRS) applies to all residential buildings, regardless of sector, in England and Wales. Whereas the Fitness Standard involved subjective interpretation with the result being a pass or fail, the HHSRS provides an evidence-based method for ranking the severity of threats posed by defects in the dwelling.[2] The system is supported by extensive reviews of the literature and by detailed analyses of statistical data on the impact of housing conditions on health. This evidence is summarised in the Hazard Profiles section of the Guidance, intended to inform professional judgement.

The HHSRS considers 29 hazards, which are related to: physiological requirements; psychological requirements; protection against infection; and protection against accidents. Each hazard is rated by reference to the 'most vulnerable age group', regardless of who, if anyone, is in occupation. This means that hazards can still be rated in properties that are not occupied. The principle is that if the property is safe and healthy for the most vulnerable age group then it will be so for people of any age.

A score is given to each hazard, reflecting: (i) the likelihood of the hazard causing harm to someone in the most vulnerable age group over the next 12 months; and (ii) the severity of the possible harm outcomes. The score enables comparison of hazards according to the type of effects (e.g. affecting physical or mental health) and whether their effects are immediate or occur over a period of time.

Based on the score given, the hazard might then be assigned a Category 1 or a Category 2 label, the former reflecting serious risks that require action to be taken by the local authority whereas judgement can be used in dealing with the latter. The enforcement options that authorities can choose from, giving consideration to the specific hazard and other circumstances, are: issuing an Improvement Notice to the landlord; serving a Prohibition Order that prevents occupancy of all or part of the dwelling; engaging contractors to carry out emergency remedial action, which the landlord must then pay for; issuing a Hazard Awareness Notice that does not require action by the landlord but informs him or her of the presence of the hazard; or using Demolition Order and Clearance Area Powers.

It is a criminal offence for the landlord not to comply with any of these enforcement orders. There is right of appeal and this is dealt with through the First-tier Tribunal (Property Chamber), though there have been relatively few appeals so it is difficult to predict how the courts will rule.

One possibility is that the form of enforcement that was issued by the local authority can be substituted. For example, an Improvement Notice given by the London Borough of Camden in March 2009 to a landlord (Mr Jones) was substituted by a Prohibition Order until the necessary works were completed. Another appeal illustrates the complexity of assessing issues relating to cold. In an appeal involving Mr and Mrs Lindsay-Taylor and the London Borough of Camden, a Category 1 Excess Cold hazard was identified and an Improvement Notice was given for central heating and double-glazing to be installed. The Tribunal found that only central heating was required.

Aside from the primary function of the HHSRS, it can contribute to assessing the health impact of defects in housing and the gains that can be made as a result of remedial work. In 2012, the BRE launched an online Housing Health Cost Calculator (HHCC) on which HHSRS data can be uploaded and stored, allowing calculation of the health costs of each hazard in a dwelling as well as for the wider housing stock.

Codes of practice

In England and Wales, local authorities are required by statute (section 167(1) of the Local Government & Housing Act 1989) to assess and publish performance indicators. Similar provisions applied in Scotland under section 17A of the Housing (Scotland) Act 1987 and section 153 of the Leasehold Reform, Housing and Urban Development Act 1993. Section 17A required local authorities annually to publish their own housing management standards and to publish how they were performing against those standards. This was repealed by the Housing (Scotland) Act 2010, which does not contain a direct equivalent. The 2010 Act brought local authorities under the control of the Scottish Housing Regulator. Authorities must now comply with the standards set by the Regulator under section 32 of the Housing (Scotland) Act 2010.

The detailed content required of these performance indicators varies within the UK, each set of requirements being drawn up by the relevant government department (Department for Communities and Local Government, Welsh Government and Scottish Government). There is a common thread, however, in that each authority reports the priority levels, the target response times for each, and the achievements of that authority in meeting those targets.

Housing associations are not included in these statutory provisions, but in England, Wales and N. Ireland, they are expected to publish similar information under the Tenants' Guarantee or according to the requirements of the Scottish Housing Regulator.

Published priorities from both local authorities and housing associations give indicators of priority categories and performance targets.

Tenants' Guarantee

The Tenants' Guarantee applies to housing associations in England and Wales. Although slightly different in each part of the UK, the general thrust is the same. Housing associations are required to meet their statutory and contractual obligations to keep their housing fit for human habitation and to ensure it is well maintained. They are expected to inspect the properties regularly and plan for future maintenance.

Tenants' Right to Repair

Reference should also be made to the law and guidance on the Tenants' Right to Repair. This legislation is intended to allow tenants to claim compensation for the landlord's failure to satisfactorily execute repairs within a given target period. The periods are set out in the Secure Tenants of Local Authorities (Right to Repair) Regulations 1994 for England and Wales and the Secure Tenants (Right to Repair) (Scotland) Regulations 1994. These provisions came into effect on 1 April 1994 in England and Wales and on 1 October 1994 in Scotland.

Although the compensation rates are different and some classes of work do not appear in both Regulations, there is a general similarity. The Regulations set out periods of working days within which the authority is required to complete the repair. If this is not achieved, the tenant can require the landlord to send a second contractor and the allowed period is repeated. Failure to complete the repair by the expiry of this second period gives the tenant a right to compensation payable by the landlord.

Both the Homes and Communites Agency (in England) and the Scottish Housing Regulator have introduced similar voluntary schemes.

The result is that for the tenants of local authorities and housing associations, there is a published target for repair of specific defects. The accumulated period between report of a defect and the right to compensation is a useful indicator of the priority attached to any given repair and the overall target period that ought to be achieved.

Although these requirements relate to the social housing sector, they were initiated in a climate of perceived lack of accountability of public sector landlords and intended to introduce a private sector competitive ethos into housing management. They serve, therefore, as a useful additional indicator of repair priorities with a more general application.

Chartered Institute of Housing standards

The Chartered Institute of Housing launched its *Housing Management Standards Manual* in 1994 and also established a Good Practice Unit. Together these provided useful benchmarks. The manual was updated regularly and described the process of identifying varying priorities in the response to disrepair and the suggested timescales for examples of defects. For instance, Table 7.1 is based on the examples of good practice in the early editions of the *Housing Management Standards Manual*.

Planned maintenance

Planned maintenance falls into two categories: cyclical and programmed maintenance. Cyclical work is carried out at regular intervals to minimise

Table 7.1 Repair completion targets for a reasonable landlord

16 View Road, VW7 8AS

Ref.	Type		Time from date of report (weeks)		
			Inspect	Order	Complete
A	**Urgent repairs**	Defective ball valve			1
		Leaking radiator			
B1	**Normal or routine repairs not requiring inspection**	Leaking guttering			4
		Windows that cannot be opened safely			
B2	**Normal or routine repairs requiring inspection**	Dampness to plasterwork	2	3	7
		Fungal decay to timbers			
C1	**Planned repairs not requiring inspection**	Rewiring			10
		Damaged fencing			
		Broken bath panel			
C2	**Planned repairs requiring inspection**	Defective floorboards	6	8	18
		Overhaul windows			
E	**Emergency repair** (i.e. serious damage to the building; danger to health; risk to safety; risk of loss or damage to occupier's property)	Falling masonry	Immediate	Immediate	1 day
		Flooding			
		Complete failure of electrics			

the risk of premature deterioration; e.g. external painting. Programmed maintenance is work required to replace components; e.g. boilers, roof coverings, windows. Usually it is work carried out to a group of buildings as a modernisation scheme.

The priority for a repair cannot be assessed in isolation as though a particular dwelling that requires attention is the sole property within the landlord's portfolio. A balance must be struck between the overall repairing obligation of the landlord and the individual demands of the tenant of one dwelling.

Responsible landlords will have a plan that allows for cyclical and planned maintenance works to be carried out to their stock. The basic works will comprise external decorations and minor consequential repairs such as repointing, renewal of glazing putties and mastic, cleaning and repair of gutters, and repairs to external doors and windows. They will also include renewal of components with a known life expectancy. The majority of such components will be parts of the service installations together with kitchen units. In some cases, planned maintenance also includes works which could be classed as improvements but which involve such small cost that they do not need to form part of a major scheme of renovation. Take, for example, maintenance of extract fans or powered ventilation systems to combat and minimise the effects of condensation.

Wholesale redevelopment or rehabilitation

The absence of new housing, the lack of availability of building land, the restriction on resources and the unexpectedly high costs of maintenance of some of the public sector housing stock built between 1955 and 1975 resulted in political pressure for wholesale modernisation schemes. The response came in the form of Estate Action Programmes, City Challenge or other one-off funding opportunities. In the context of disrepair and repair timescales, these schemes added another generally imprecise variable.

It is reasonable to say that a landlord should not be obliged to carry out expensive and substantial works of repair when, within a relatively short period, all that work will be undone by a major renovation scheme. The difficulty arises primarily in the funding system and this is perhaps best illustrated by an example (see Box 7.1).

A major improvement programme will usually take several years to complete. But the authority will only know what capital allocation it will receive one year at a time, and this allocation is not 'ring fenced' to a particular scheme. For example, a major improvement scheme can be planned out over three years. In year one, enough allocation will be received to cover the plan for that year. In year two, a smaller overall allocation may be received because of the decision-making process outlined above. Meanwhile, some other estate could suddenly become a higher priority,

Box 7.1 Example

A local housing authority owns a large 1960s housing estate, including some high-rise blocks, built of nontraditional construction. The major problems on the estate are manifested in social unrest and high unemployment. The infrastructure suffers vandalism, and the fabric of the dwellings is degraded and most suffer from dampness. The external structure of the high-rise blocks is failing, allowing water penetration; and services such as lifts are at the end of their useful life.

The landlord authority recognises that responsive repairs are running at a very high level and costs are unusually high but also that serious defects still remain. It also recognises that the works really required will be major and of high capital cost. Access to funding for such works is primarily from central government, although this may be a 'leverage exercise' to obtain additional funding from the private sector.

Government funding schemes seem to rarely work on a predictable or planned basis for the recipients. City Challenge, for example, was in effect a lottery, where competing bids from local authorities were assessed and only a few lucky winners got money. Those that did win got money for five years and were clearly in a much better position to plan and carry out major estate improvement programmes.

Other sources of government funding seem more problematic. Those seeking funds are often perceived as exaggerating the costs and/or need, and those controlling the funds see their role as one of reducing the demand to a level which the supply can accommodate. Decisions about how much the authority can borrow are based less and less on the immediate needs for repair and improvement and more on other performance indicators such as: how well the authority fulfils its enabling role; the quality of its overall strategy; and its housing management performance. Thus authorities with the very pressing need to deal with estates like the one described above can easily get less money than those with far fewer problems.

perhaps where a fire has revealed some potentially fatal design defect that has to be remedied immediately. The combination of these factors is likely to mean that the authority has to push much of its programme for the original estate from year two to year three.

The overall effect of this is that a local authority could, for example, rightly say that a dwelling should be included in a major renovation scheme within the next year or three years, thus lowering the standard of repair required and affecting the priority of some repair works. The occupants may reasonably be expected to endure conditions which will be overcome in a couple of years, but what if funding does not arrive and the works are not put in hand after, say, three years? The landlord may still assert that the renovation scheme will occur within two or three years. Even if this does happen, this would mean that the occupants have endured the conditions not for two years as originally anticipated, but for five or six years. The period could of course be longer. The anticipated life of the dwelling which was, at the beginning, two years or so may well stretch, in reality, to ten years.

Box 7.2 Points of note in relation to the redevelopment of commercial properties

Useful guidance on the application of major schemes to the standard of repair can be gleaned from the commercial world. Under section 18 of the Landlord and Tenant Act 1927, a landlord's right to damages for disrepair (dilapidations) can be removed where the landlord intends to redevelop. The intention to redevelop must however be tested, primarily by the outgoing, and therefore liable, tenant. It must be more than a contemplation of redevelopment and it must be that the landlord has a reasonable prospect of carrying out the plan (*Cunliffe* v *Goodman* [1950]). The consideration of whether there is a reasonable prospect of carrying out the redevelopment will include the availability of funds.

The inclusion of the availability of funds to assess the validity of a landlord's intention to redevelop again arises in the commercial world in the renewal of leases. Under the Landlord and Tenant Act 1927, lessees enjoy a right to renew their lease on expiry. The landlord can object to this renewal on several grounds, one of which is the intention to redevelop. In contrast to the position under section 18, the onus falls on the landlord to demonstrate the validity of the claim, rather than on the tenant to show the landlord's intentions. Again, the availability of funds has been considered a relevant factor.

As with so many matters, these considerations must be assessed on the merits of each case, using guidance and expert opinion where appropriate. There is no easy code to follow to arrive at the effect of a possible, and even desirable, renovation scheme on the timescale for response to repairs.

Value for money

Where, for example, windows are made of timber and some, but not all, are rotten and require repair, the most cost-effective repair would be renewal. If the dwelling is one of a number with similar defects, the obvious way to maximise value for money is to place an order for a complete window replacement. What is the landlord to do? The tenant should not be expected to endure rotten windows for an unknown period. The dwelling demands a standard of repair which will include sound and well-repaired windows. Yet, the landlord must balance these considerations against the value for money criterion. If a firm date for a window renewal programme is available, the courts are likely to be sympathetic to the landlord deferring repair. However, in the absence of precise dates for such works, it is likely that a repair of the window could be ordered as a result of legal action.

The repair ordered may not be renewal, but as suggested above, renewal may be a better option. The landlord can, therefore, be forced to execute uneconomic repairs which will be undone before their life expectancy has expired.

Another value for money consideration will have the effect of bringing forward a repair that may otherwise be deferred. Where access equipment, such as scaffolding, is required to carry out some works that are taking place sooner than the particular repair being considered but that same access will be required to execute the repair in question, it makes economic sense for the landlord to execute both repairs together.

An example would be the repointing of a chimney stack. Defective pointing is a defect which degrades relatively slowly. It would normally be left until a convenient time when access is readily available. If, however, a chimney pot has fallen off – possibly blown off in a storm – and requires urgent repair, similar access will be required. The extra cost of executing the repointing as part of the reinstatement of the chimney pot will probably be marginal, so it would make sense to execute both together.

The desirability of executing a repair earlier is not likely to persuade a court to order a shorter period for that repair, whereas deferring a repair for a period in the interests of value for money has been accepted as reasonable. The length of deferment is one of consideration on merits and expert opinion (see Chapter 9).

Further recommended sources of information include:

Using the Housing Act 2004, by Helen Carr, Stephen Cottle and David Ormandy, 2008

First Tribunal/Residential Property Tribunals website: https://www.justice.gov.uk/tribunals/residential-property/decisions

Building Research Establishment Housing Health Cost Calculator: see https://www.housinghealthcosts.org

Good Housing Leads To Good Health – a toolkit for environmental health practitioners published by the Chartered Institute of Environmental Health (CIEH): www.cieh.org/uploadedFiles/Core/Policy/Housing/Good_Housing_Leads_to_Good_Health_2008.pdf

Notes

1 Department for Transport, Local Government and the Regions (2002) *Housing Disrepair Legal Obligations: Good Practice Guidance*, London: DTLG. Available at: https://www.gov.uk/government/uploads/system/uploads/attachment_data/file/7852/142949.pdf
2 *Operating Guidance*, dated February 2006, can be downloaded from: https://www.gov.uk/government/publications/hhsrs-operating-guidance-housing-act-2004-guidance-about-inspections-and-assessment-of-hazards-given-under-section-9

Chapter 8

Follow-up action

It is central to the effective management of properties that essential repairs are executed at the appropriate time. In most cases where a prompt response is not achieved, this is due to a combination of factors. The housing manager may lack time, expertise or assertiveness and may be satisfied that once the works are ordered, his or her own responsibility ceases. Those responsible for carrying out the works may set priorities not by the need for the works to be executed but by the availability of skilled workers and materials. The use of external contractors can pose the same problems and, in many cases, increase the diversity of conflicting priorities.

Monitoring performance

Many social housing landlords assess and publish the relative performance of different contractors. As we see in Chapter 7, performance measurement is a statutory or regulatory requirement and assessment of the performance of individual contractors is an integral part of this. These requirements also oblige local authorities and housing associations to have much better systems for tracking repairs and, in particular, to be able to pinpoint areas where they are underachieving; for example, by geographic location, by contractor and by priority band.

Nonetheless, the basic ingredient which remains essential is the effective monitoring of the whole process by the housing manager.

Management systems

The housing manager requires tools to assist in performance assessment. A visible and transparent audit trail is an essential tool to demonstrate actions, both considered and taken. Although the tools could be manual, electronic media should be used to maximise effectiveness. These might comprise an electronic system of records linked to a database and diary which automatically notifies the housing manager of the approach of a deadline for completion and the anticipated date of execution as well as generating correspondence to ensure that times are achieved.

It is not the function of this book to describe, to analyse or to design such systems, but the principle demands that the housing manager should make of such a system can be considered. The key pieces of information are:

- when the order is placed;
- when it is to be completed;
- when it needs to start in order to achieve completion;
- when it starts;
- when it is completed; and
- if delayed, why works are delayed and the revised dates for start and completion of works.

By monitoring the execution of the works, the housing manager is able to ensure that target times are kept.

Tenant satisfaction surveys

Increasingly, social housing landlords are asking their tenants to complete satisfaction forms. These are a useful tool not just for producing statistics of customer satisfaction but for indicating both successes and deficiencies in the repairs service. However, tenants do not always complete the forms, and logging responses absorbs resources that often cannot be spared.

Positive responses that confirm all works are satisfactory and completed within the target time contribute to the statistical audit. The forms that express dissatisfaction either with the works or with the time taken should be cross-checked against the positive responses. In addition, returned forms that suggest by their date that works were completed late should generate a cross-check.

With major works, there is a supervisor and a contract administrator responsible to their landlord client for confirming that works are satisfactorily completed and that payment can be made. A record of this note of satisfaction should be available to the housing manager.

Perhaps the most important source of information on the adequacy of repairs is any subsequent complaint by the tenant. Complaint records vary widely, but the design of some systems meant that relevant information was concealed. For example, the design of one local authority's computerised records system meant that complaints were overwritten with each modification. The effect of this was that after repairs were completed, there was no readily accessible record to show that the defect associated with the original complaint had in fact been rectified.

These days, however, computerised databases are becoming more and more sophisticated, and integrated repairs and ordering systems are in place in all local authority and housing association offices. In these systems, everything relating to a particular property is accessible through the database and, conversely, no information that is recorded about the property can be removed. Each piece of information is identifiable so that the date and source of each entry can be traced. Overwriting can be prevented and records are kept cumulatively. Advances in program design are rapid and systems are offering greater support to housing managers; for example, generating a daily report showing repairs that are outstanding beyond the target date.

Redecoration

Additional to the repairs is the question of consequential works. In housing disrepair, such consequential works primarily involve redecoration, which traditionally has been seen as the tenant's responsibility. If redecoration is necessary following execution of works by the landlord then, commonly, this is either left for the tenant to do or a minimal contribution is made towards materials on the basis that the tenant intends to redecorate in any event.

Neither of these solutions is attractive for the tenant. With no contribution, the landlord may repair the defect but not restore decorations to the condition they would have been in had the defect not occurred – at least in the tenant's view. Conversely, the landlord may take the view that if the decorations were poor prior to the defect then there is no consequential loss. Assistance in the discussion of this aspect can be found in the cases of *McGreal* v *Wake* (1984) and *Bradley* v *Chorley BC* [1985].

Variations of the works

Changes may occur in the required works in two ways. First, the works may have to be specified as provisional until exposure of concealed areas allows exact detailing of the repair. Second, there may be further works or consequential works arising out of the repairs.

When specifying works, it is often necessary to allow for two or more options, the selection of which will be made either solely by the operative or by a supervisor. As an example of a defect that requires a series of repair works, let's look at dampness at low level to the walls of the lowest floor of a dwelling. The defect that can be identified on inspection is the dampness to the wall. Assuming that only limited testing is carried out (see Chapter 3), the order for works needs to allow for some exposure to the wall before deciding on remedial works. This allowance for options or variations must be built into any ordering process and be reflected in the records of the housing manager. This type of defect may also require the execution of one or more repairs followed by a period of monitoring before further works are specified.

The works actually undertaken need to be recorded so that (i) they can be checked against the options envisaged, (ii) they can be referred to later should the initial works not prove wholly successful, and (iii) the experience can assist in building up a source of knowledge both in relation to the particular dwelling and in relation to the type of defect found.

Because some works are not fully identifiable at inspection stage, further major works may become apparent during the course of what are envisaged as relatively simple repairs. For example, the problem of a loose floorboard on the ground floor of a dwelling may, without intervention and inspection at an early stage, conceal a substantial outbreak of fungal decay. Having started the works, there is little option but to continue. However, the cost and scale of the works may change significantly and cause substantial unanticipated disruption to the occupants. The eradication of fungal decay may require the use of toxic materials which means that the occupants have to be temporarily housed elsewhere.

The procedures in place for the implementation of repairs and their monitoring by the housing manager must include a mechanism for such unexpected emergency situations. Equally, such situations underline the need to carry out as much investigation and pre-planning of the works as possible.

Longer term action

Initial repair works may point up deficiencies that need to be addressed, both in terms of long-term maintenance of the property's value and in terms of healthy and acceptable standards for occupation by a tenant.

In determining the need for action, many authorities and associations have used the National Home Energy Rating scheme, particularly as part of stock condition surveys. This gives a thermal insulation value for each dwelling from 0 (non-existent) to 10 (excellent). Targets can then be set to achieve, for example, a rating of 5 for all homes within three years, and of 7 within six years. Allied to this are targets for reduction in the number and cost of repairs.

One of the most common defects is condensation-caused dampness. To eradicate this defect fully, a combination of insulation, ventilation and affordable heating is required. However, the work done by the landlord may be limited to one of these three elements – probably the installation of extract fans to the kitchen and bathroom, which is the cheapest option. It may be that further works such as the installation of fixed electric heaters is also carried out, but these are of minimal effect unless the running costs are affordable for the occupants.

As explained in Chapter 3, the presence of dampness causes many problems and should be remedied, even though to do so with full effect may exceed the repairing obligation. A procedure should be in place for the identification of major works or improvements to eliminate the need for repeated repairs and to safeguard the landlord's property.

In order to fully assess the need for energy improvement works or to assess the efficiency (or lack of it) of existing heating systems, it is useful to consult Sutherland Tables, which are a reputable and established source of independent and impartial information on domestic heating costs. These tables, which are available on a subscription basis, provide comparative costs for space heating and hot water for the most common fuels across a range of standard house types throughout the UK and Ireland. The tables also provide indicators for heating an individual room using a variety of appliances including open and closed solid fuel fires, gas and electric heaters, and both fixed and portable liquid petrolem gas heaters. The tables are compiled quarterly in January, April, July and October and are available to download from www.sutherlandtables.co.uk/. Sutherland also provides archive copies which can be purchased.

Monitoring short-term repairs

Defects require monitoring where short-term works have been executed with the expectation that longer term solutions will be implemented in a known timescale, but when those solutions then cannot be or are not implemented.

The short-term solution can sometimes outlive its anticipated life. Where repairs of a known limited life expectancy are used, and imposed on the tenant, because of a forthcoming scheme of works, it is only equitable that the defect should be revisited when the circumstances change.

Visits and revisits

If repairs are not executed, or if they prove ineffective, then further inspections are required. It may be that by that stage, the matter is being handled by some other member of the housing team, perhaps a maintenance manager; nevertheless, the housing manager needs to be aware of the activity.

At each subsequent inspection, the original schedule and report should be revised to show the change in conditions. This enables the tracking of each individual defect which can prove invaluable at a later date. Again, computerised systems can assist greatly with the tracking of repairs history.

Chapter 9

Court proceedings

A court hearing can result in an order for works to be done, a money judgement of compensation payable or a legal resolution to a dispute on liability. In the course of proceedings, both the facts and the interpretation of the law may be in dispute. Both of these can affect the liability for carrying out the repair as well as any payment of compensation for the failure to repair.

Some disrepair cases may be settled at the door of the court; but nevertheless, the housing manager has to be equipped and ready to attend court and present her or his evidence to a judge. The majority of cases that housing managers deal with are tried under civil law in the county courts. Some may go the High Court, but this is rare for disrepair.

Environmental Protection Act prosecutions are dealt with in the magistrates' court under the criminal justice system. Although the standard of proof is different and the rules regarding advance disclosure of evidence are less rigorous than in civil cases, these prosecutions can be dealt with on a quasi-civil basis with advance discussions between the parties and their advisers minimising the costs of any hearing.

In many of these cases, the housing manager will be a lay witness of fact, not an *expert*. Accordingly, the same rules of conduct apply as to any other witness in the case. Witnesses of fact are traditionally excluded from the trial until after they have given evidence. However, in both criminal and civil proceedings concerning housing disrepair, all witnesses are usually allowed into court unless specific objection is made.

Instructing an expert

This decision to instruct an expert needs to be made at the stage when the complaint is first received and then reviewed at each successive stage. The following questions should be answered:

- Is there an adequately qualified/experienced staff member who can properly assess the complaints?
- Are the facts concerning the complaint clear, from both sides?

- Are the defects complained of clearly the liability of one party or the other?
- Are the required repairs clearly specified?
- Is the matter definitely going to be resolved without going to trial?

Only if all of these questions are answered in the affirmative can the appointment of an expert be excluded. The time to appoint an expert depends on when the answers to the questions above stray into the negative or even to 'don't know'.

It is false economy to allow a complaint to proceed towards litigation and trial without assessing the merits of the case.

Expert evidence and advice

The role of an expert witness is clearly distinguished from that of a lay witness. The expert, unlike the lay witness, is required not only to deal with facts known to her or him but also to express an opinion on the interpretation of those facts in the context of the case before the court. For this reason, the expert evidence presented to the court must be independent and unbiased. Guidance on this was offered in the decision of Mr Justice Cresswell in the case of *Ikarian Reefer* [1993] 2 Lloyd's Rep 68.

The duties and responsibilities of expert witnesses can be summarised as follows. An expert witness should:

- present evidence to the court that is independent and not influenced by the demands of litigation;
- provide independent assistance to the court by giving objective opinion on issues within her or his expertise;
- never take on the role of advocate;
- state the facts or substance on which her or his opinion is based;
- not omit to consider material facts which detract from the conclusion;
- make it clear when a particular question falls outside of her or his expertise; and
- state her or his opinion as provisional if insufficient research has been possible.

Expert advice, on the other hand, is opinion based on the facts presented, which is given to the party on whose behalf the expert is instructed. This advice may include matters that are potentially damaging to the client's case.

The case of the *Ikarian Reefer* is still the definitive case in respect of the duties and roles of an expert witness, and the introduction of the Civil Procedure Rules in 1999 was designed, in part, to reinforce that. In 2000, His Honour Judge Toulmin further refined the definition in *Anglo Group plc v Winther Brown & Co Ltd*.

The role of experts is now governed by myriad rules. The First-tier Tribunal dealing with HHSRS appeals does not have rules requiring the same approach but will expect a similar approach to be followed (see Chapter 7).

For England and Wales, the framework within which experts must act is provided by Civil Procedure Rules Part 35 – experts and assessors (CPR 35) and the accompanying Practice Direction (PD 35).

From 2014, help to interpret and comply with CPR 35 and PD 35 is provided in *Guidance for the Instruction of Experts in Civil Claims*. (This replaced *The Protocol for the Instruction of Experts to Give Evidence in Civil Claims*.)

Yet further guidance is to be found in the *Pre-Action Protocol for Housing Disrepair Cases*, which aims to have problems resolved before going to court. This applies to claims, not counterclaims, but the general principles set out represent good practice.

For surveyor experts in disrepair, there is also the guidance published by the Royal Institution of Chartered Surveyors to be borne in mind. This includes:

- *Surveyors Acting as Expert Witnesses*, 4th edition (2014);
- *Dilapidations*, 6th edition (2012);
- *Dilapidations in Scotland*, 2nd edition (2011).

The guidance set out in these publications can, of course, apply to other professions also.

The line between the role of impartial expert and the obligation to those who give instructions is a delicate one and can be very hard for the expert to tread. However, the duties imposed upon expert witnesses mean that any damaging factors must be taken into account in the presentation of evidence to the court (see also Box 9.1 for discussion of the possible effect of remuneration on the role of expert witnesses). In some well-publicised cases, experts have been criticised for straying from their expert witness role.

In *London & Leeds Estates* v *Paribas* [1993], the expert was ordered by the court to disclose a proof of evidence from another case, which went to arbitration, on the basis that his approach was inconsistent. The court stated that an expert must resist any subconscious tendency to become a member of a team and, as a result, to proffer views unduly favourable to the position of the party instructing him.

In *Kenning* v *Eve Construction* [1995], the expert had written a report for disclosure and in a covering letter expressed views contrary to his client's case. The judge concluded that the party instructing the expert could either not call the expert at trial, and therefore exclude *all* his evidence, or call the expert, in which case the whole of the expert's opinion should be disclosed, damaging or not to the client's case.

Choice of expert

In cases of housing disrepair, the criteria that should be the basis for selection of an expert include proven technical and legal experience.

A thorough knowledge of the type of building under consideration is required together with a good understanding of the relevant laws so that the opinions expressed on liability for a defect and/or the extent of repair can be accurate.

Previous experience as an expert witness is a further criterion for selection. A successful track record adds to the expert's stature in the eyes of the court and allows the instructing party to have confidence in the opinions expressed. The expert should have a record of being able to substantiate opinions and reporting at trial. For this reason, the expert may, on occasion, need to play devil's advocate in the investigative stage of preparing the report so as to tease out facts. For example, there may have been an inexcusable delay in carrying out a minor repair which would have prevented a greater defect. For example, the running overflow from a defective ball valve – requiring a simple and cheap repair – if ignored, will cause dampness to the structure and possible fungal decay.

There will, of course, always be cases where facts are disclosed only at a late stage – sometimes even during the trial – and these will affect the opinions expressed. For example, in cases of dampness where the contribution of condensation and penetrating damp to the overall damp conditions must be assessed, the disclosure for the first time at trial by the tenant that a flueless gas heater has been used may significantly alter the balance between two competing explanations.

Box 9.1 Fees for expert witnesses

Attention has been given to the issue of expert witnesses in housing disrepair cases as part of the reform of the legal aid system in England and Wales. The priority for government has been to limit spending on fees, focusing resources on cases that genuinely require expert input as well as ensuring that support is provided fairly. From the perspective of those acting as expert witnesses, however, there are several issues that need to be considered when determining the level of remuneration.

Back in 2008/9, the Legal Services Commission had started looking at experts' fees in housing disrepair cases, the outcome of which was promulgated in 2011. The assessment was based on the cost of a very basic survey and the rate was eventually set at £50 per hour. Not surprisingly, those of us acting in housing disrepair matters took a dim view of this. During the earlier consultation period, experts – many of whom were surveyors – were asked whether they would be prepared to act if fees were reduced. The answer in most cases was in the negative. The Ministry of Justice failed to heed this warning and was ill-prepared for the difficulties faced by lawyers/advisers in engaging experts after the fees had been restricted.

Following up on this initial work – and with input from the Housing Law Practitioners Association, Shelter, the Advice Services Alliance and the Legal Aid Practitioners Group – the Ministry of Justice has now concluded its review, noting the average hourly rates for surveyors in disrepair cases as follows: £125 per hour in London and £95 per hour out of London. As rates for all experts (alongside solicitor and counsel's rates) were to be codified and reduced by the Ministry of Justice by an average of 10 per cent, the new guideline hourly rates for surveyors in housing disrepair cases have been settled as: up to £115 per hour in London and up to £85 per hour outside London (the rate is apparently determined by the location of the expert, not of the subject premises). This rate is now confirmed in the Civil Legal Aid (Remuneration) Regulations 2013, which came into force on 1 April 2013.

Although things are moving in a better direction in terms of levels of remuneration, another difficulty for expert witnesses is that detailed cost quotations are now being sought from experts who must estimate in advance the time to be spent on inspections, etc. This is not always easy, based as it will be on sparse instructions. The temptation is to overestimate. Some experts work on a maximum fee basis rather than a quotation.

Another oddity is the apparent conflict with the Ministry of Justice Civil Procedures Rules Part 35 (Experts and Assessors) (CPR 35), in particular 35.8 (3)(a), where fees for a single joint expert have been customarily paid equally by the parties. A party that receives legal aid is limited to the rates set by the Ministry; but where an expert has to subsidise this by charging a higher hourly rate to the privately funded party then, if not specifically agreed by the parties and the court, this could potentially be contrary to the custom of CPR 35.

Housing manager's evidence

Witness statements by both sides are exchanged before trial. At trial, these statements will form the heart of the evidence given by each witness.

Keeping proper records and tracking events should mean that an effective statement can be prepared on the basis of that information. The statement needs to rehearse the history of the case from the first complaint or notification of a defect through to the present stage. Later, a final statement may be confined to a particular area or time. Information for this is drawn from the notes, reports and photographs prepared by the housing manager as well as records held on housing files. Photographs showing the general conditions assist the court to form a view of the overall condition of the property. Close-ups of individual defects are often difficult to show and may require a higher level of photographic skill than is available.

Compliance with orders

Any works to be executed as part of a court order should be executed methodically (see Chapter 8). Where genuine reasons cause delay in complying with a court order – for example, unforeseen works or refusal of access by the tenant – then the housing manager must be alert to the need either to agree with the tenant that the time allowed in the court order can be extended or be prepared to return to court to obtain an extension.

If a return to court is necessary, the housing manager must attend fully equipped with all the reasons to justify the application. This may include letters and reports as well as personal evidence of the reasons for delay. Any extension of time should be sought prior to expiry.

Chapter 10

Conclusion

Achieving good value from the repairs executed can improve tenant satisfaction, reduce the pressures on housing managers, save money for the landlord and save on resources. To achieve good value, defects must be correctly diagnosed, the remedy identified and the remedial works executed in accordance with a pre-planned priority.

Procedures

The application of the ISO 9001 family of standards for Quality Management to housing management is still not common, but coupled with the performance criteria for housing set down by monitoring bodies and by statute, this brings a real need to establish written procedures. These need to be monitored both in terms of their implementation and the extent to which they work in practice. To prepare a procedure untried and untested is a wasteful and often destructive activity. Procedures, at least in the context of ISO 9001, need to be relevant and have operational credibility.

The inability to adhere to procedures generates hybrid methods and encourages formal and informal ways of doing work. This duplication brings the strategic objective of a quality service that meets customers' needs into disrepute and defeats the aim of learning from continuous review and improvement of operational processes. The implementation of new procedures should be preceded by a thorough audit of existing operations actually carried out by existing staff. Long-serving staff should be involved in the consultation.

Efficient use of resources

By understanding that a defect is not necessarily only what can be seen, the housing manager will be able to assess whether what is visible is a symptom of a more serious or less serious defect than it first appears. This understanding enables informed judgements to be made on the type of repair required, its priority, and the implications for both the landlord and the occupants. The housing manager can more accurately relate the circumstances to the legal obligations of the landlord. This will enable the setting of priorities and monitoring of progress with an eye not just to the immediate efficient use of the repairs resources, but also to minimising expenditure of the limited funds on litigation.

To end this concluding chapter, I return to the issue of remuneration for expert witnesses, which was raised in Box 9.1. This ties in with issues concerning the recruitment, training and retention of professionals in housing disrepair. In terms of those acting as experts, individuals with a limited sphere of technical knowledge or professional experience may be an answer to any supply shortage. The downside is that at the beginning of a case it is not always possible to be certain of how narrow the expertise required will be. Where there is certainty of scope, a restricted skill set may work; but where, as is often the case, the expertise needs to draw on a wider range of knowledge and experience, limited, narrowly focused experts may not be able to deliver what is required for the parties or for the court.

Over the past 10 to 15 years, entries to the surveying and engineering professions have declined as those to other professions such as the law and accountancy have increased. As so many of us approach the end of our working lives, is it not now time to ensure that we all encourage the young to see these roles in the profession as interesting and rewarding career moves? This, however, requires a perceived level of remuneration comparable with other professions. The government must be mindful of this in the broader arena of career choices, not just in respect of litigation.

Without expert involvement, the burden falls back on the parties and on the court. The parties are often unable to make findings which could assist the court, and misdiagnosis of the cause(s) of a defect are not uncommon. This will result in wasted resources and prolonged dissatisfaction. The court is faced with mounting pressures and judges are no better equipped to determine technical issues.

Appendix 1
Building diagrams

Appendix I

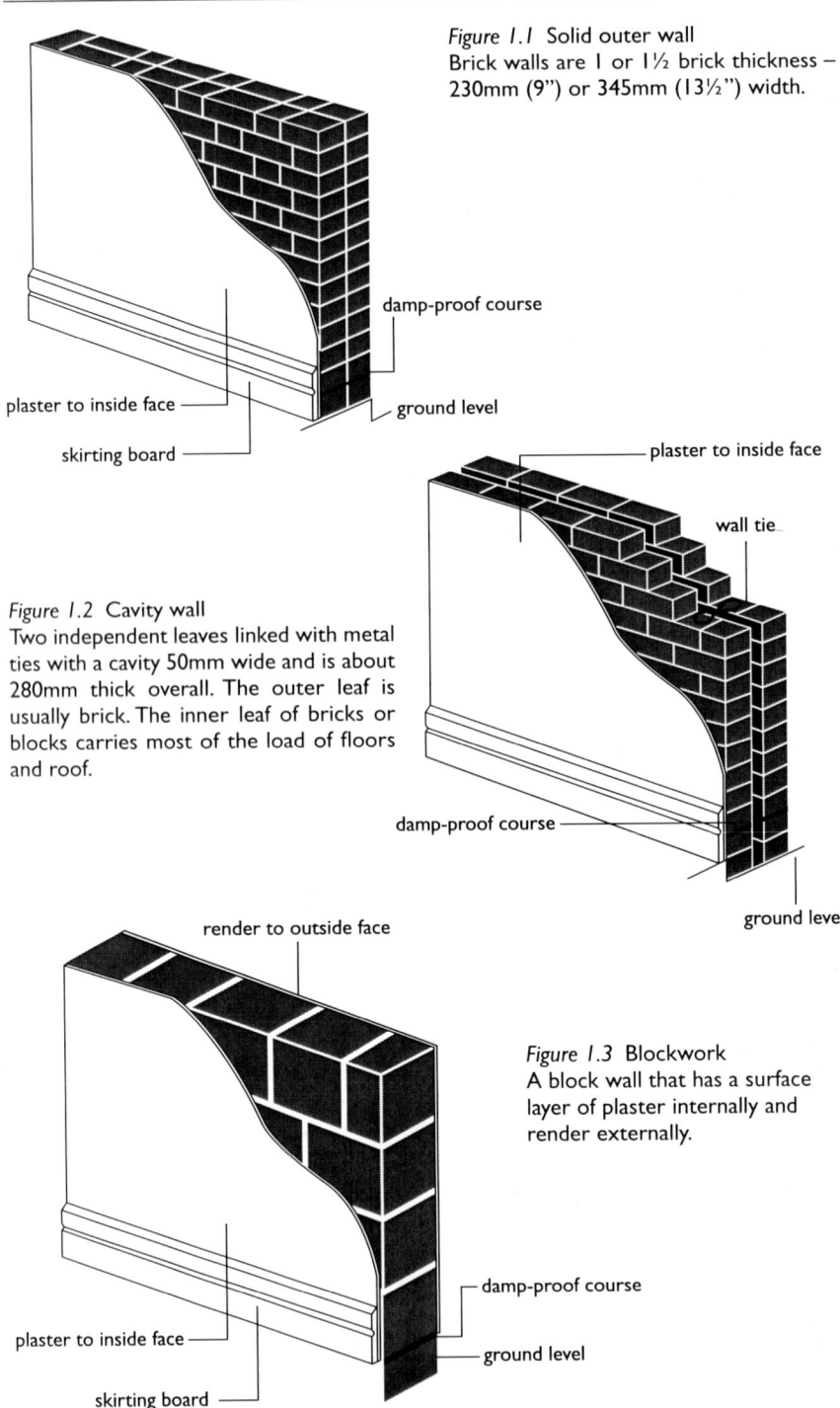

Figure 1.1 Solid outer wall
Brick walls are 1 or 1½ brick thickness – 230mm (9") or 345mm (13½") width.

Figure 1.2 Cavity wall
Two independent leaves linked with metal ties with a cavity 50mm wide and is about 280mm thick overall. The outer leaf is usually brick. The inner leaf of bricks or blocks carries most of the load of floors and roof.

Figure 1.3 Blockwork
A block wall that has a surface layer of plaster internally and render externally.

Appendix 1   107

Figure 2.1 Timber stud partition

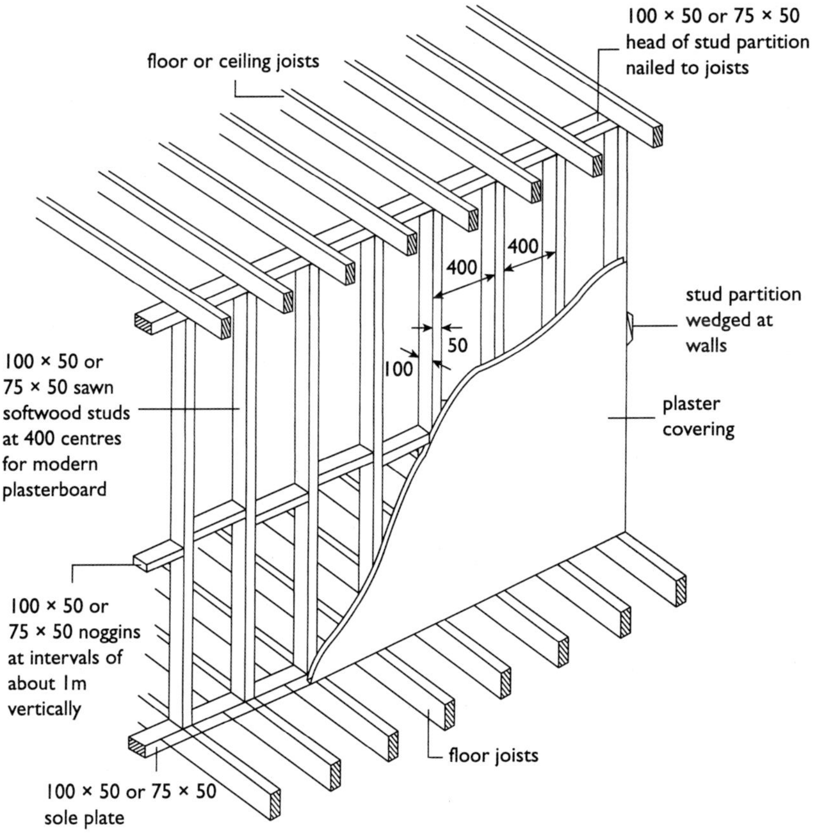

100 × 50 or 75 × 50 head of stud partition nailed to joists

floor or ceiling joists

stud partition wedged at walls

plaster covering

100 × 50 or 75 × 50 sawn softwood studs at 400 centres for modern plasterboard

100 × 50 or 75 × 50 noggins at intervals of about 1m vertically

floor joists

100 × 50 or 75 × 50 sole plate

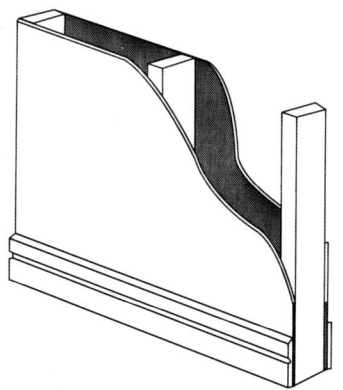

Figure 2.1 (a) Stud partitioning
This common interior wall is made from a framework of timber studs. Today it is faced with plasterboard. The wall may, or may not, be load-bearing.

Figure 2.1 (b) Lath-and-plaster wall
Found in older houses. The plaster is about 25mm thick and mixed with horsehair to increase its strength. The plaster is bonded to horizontal laths of split timber nailed to the timber uprights. If the wall is load-bearing, it may have diagonal struts between the timber uprights.

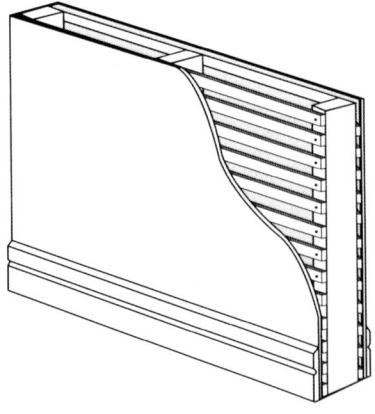

Figure 2.2 Blockwork partition
A block wall that has a surface layer of plaster.

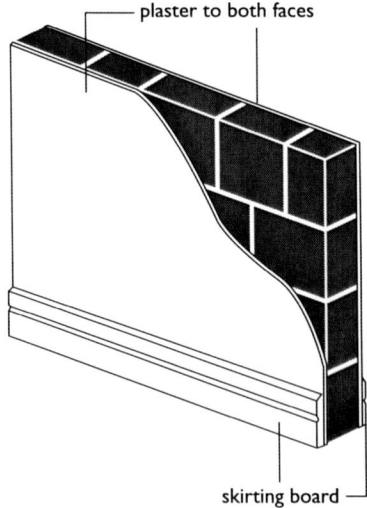

Appendix 1   109

Figure 3.1 Timber plates

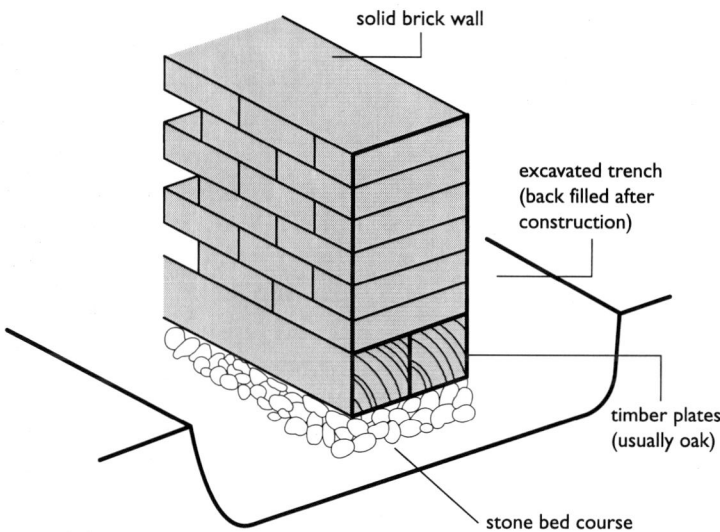

Figure 3.2 Brick stepped footings

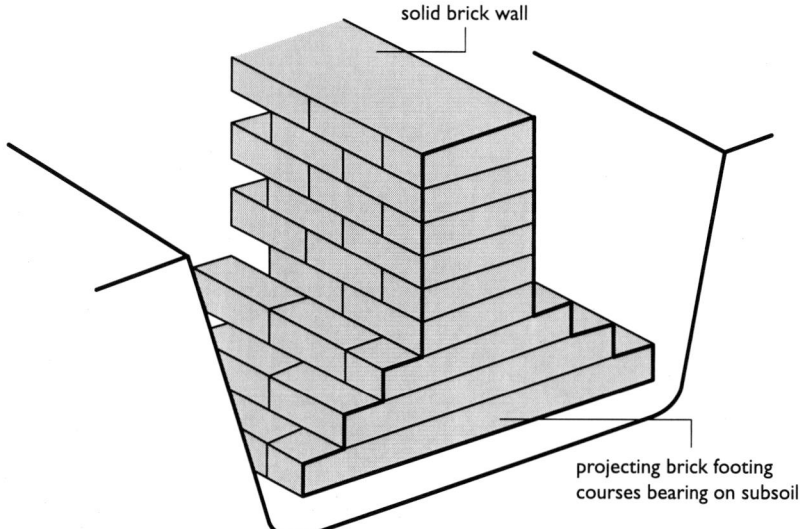

Appendix 1

Figure 3.3 (a) Concrete strip foundations

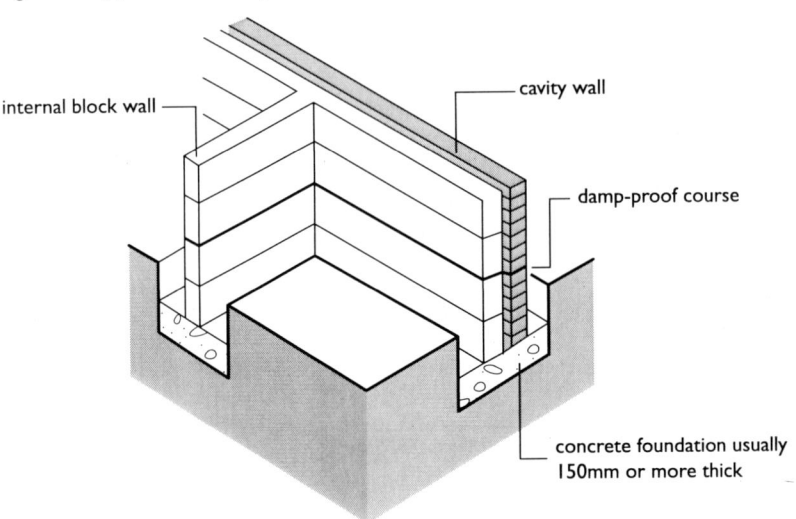

Figure 3.3 (b) Narrow strip foundations

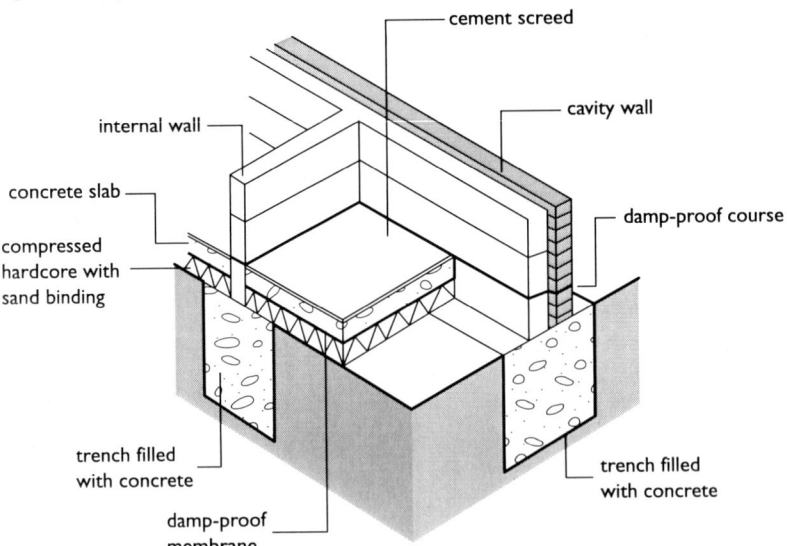

Appendix 1

Figure 3.4 (a) Flat raft foundations

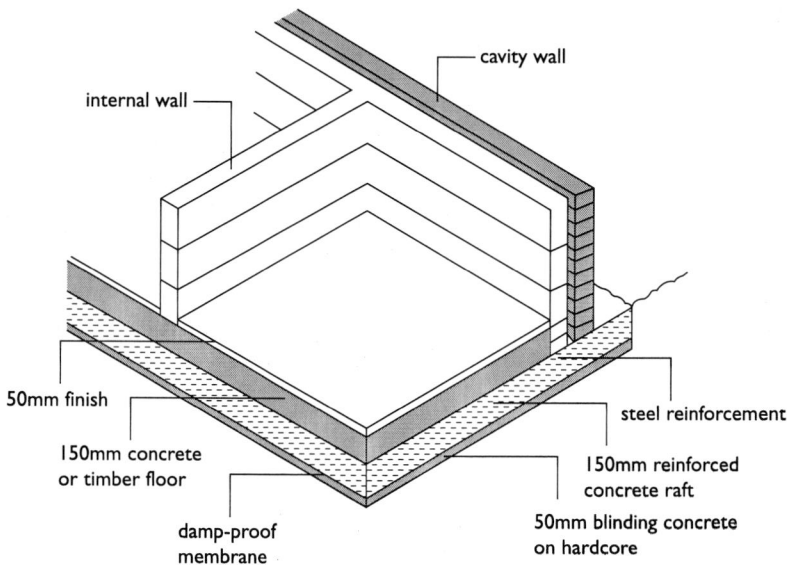

Figure 3.4 (b) Wide toe raft foundations

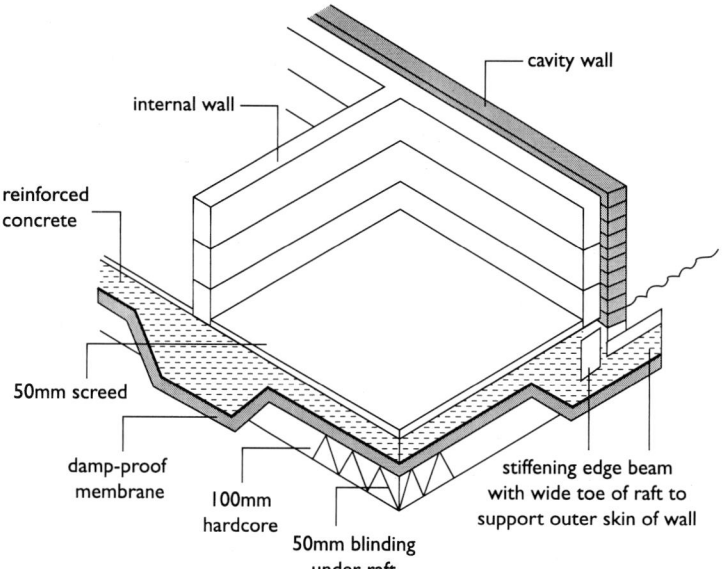

Figure 3.5 Pile foundation

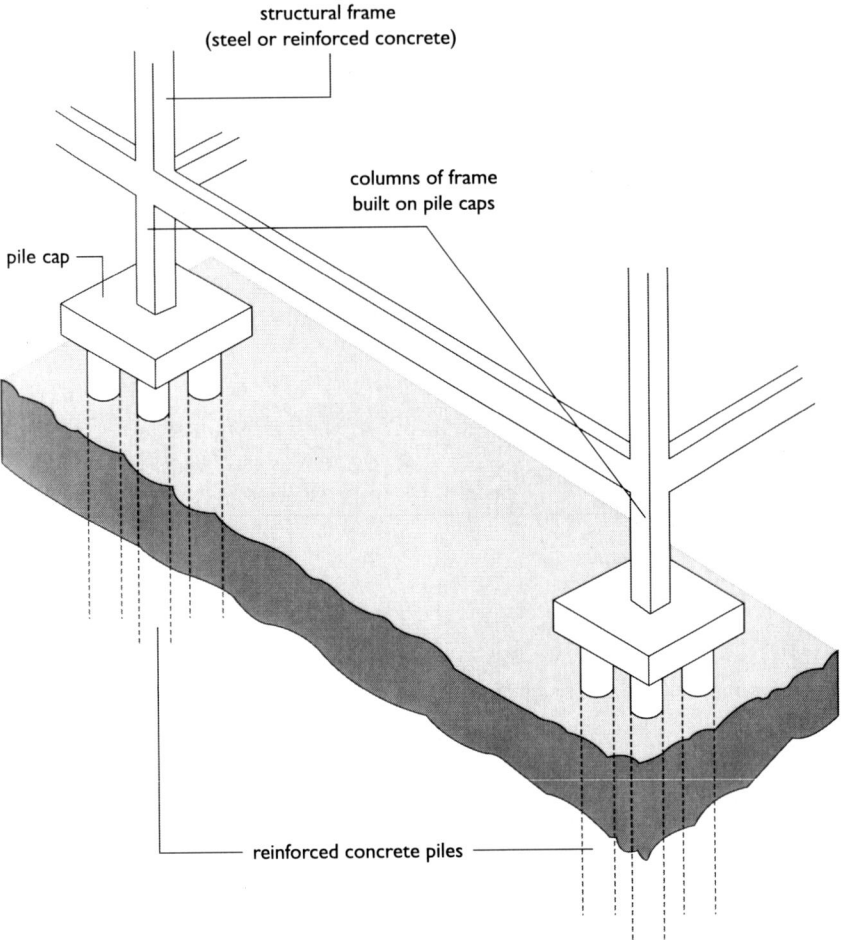

Figure 4.1 Pitched roof

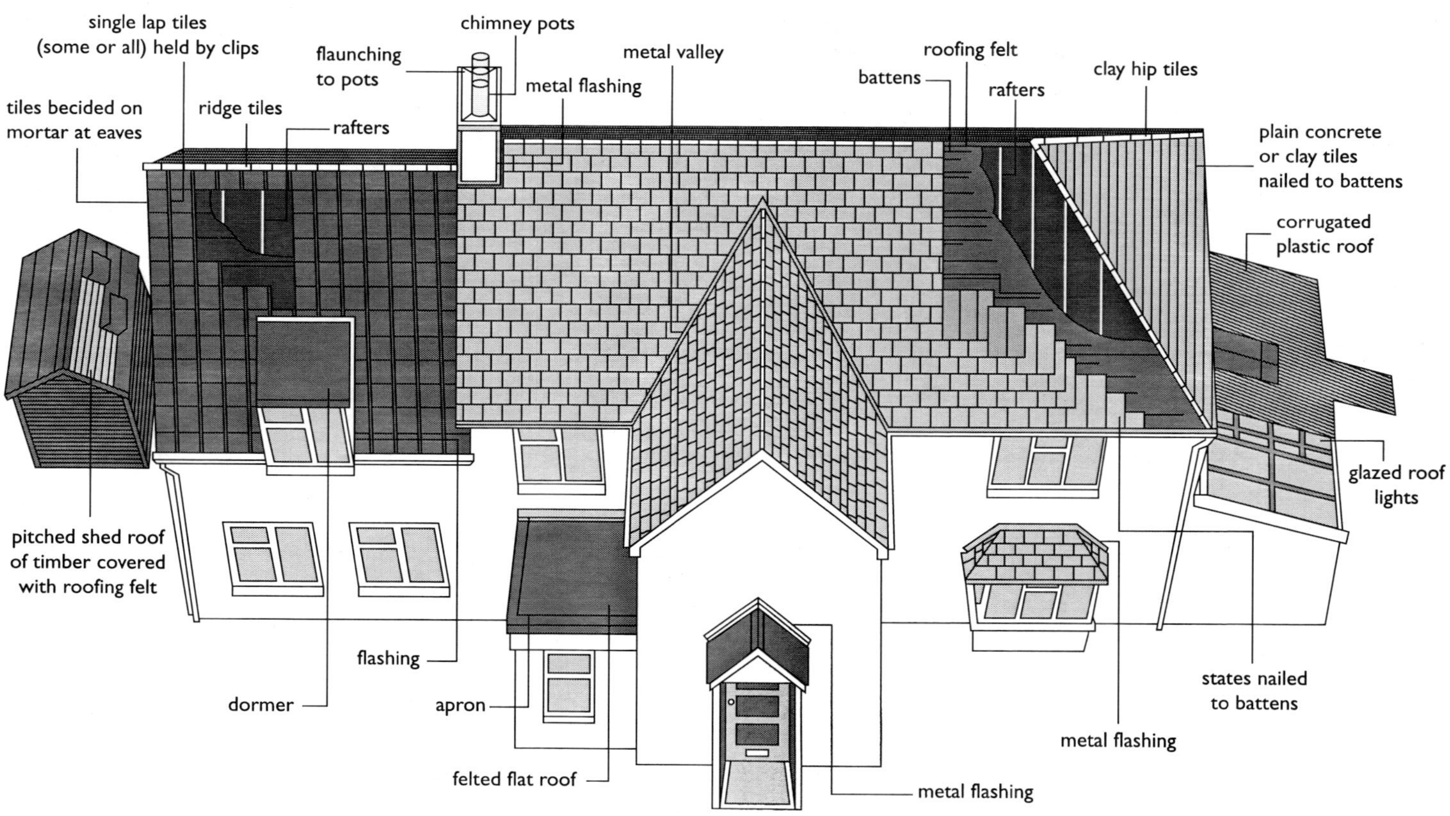

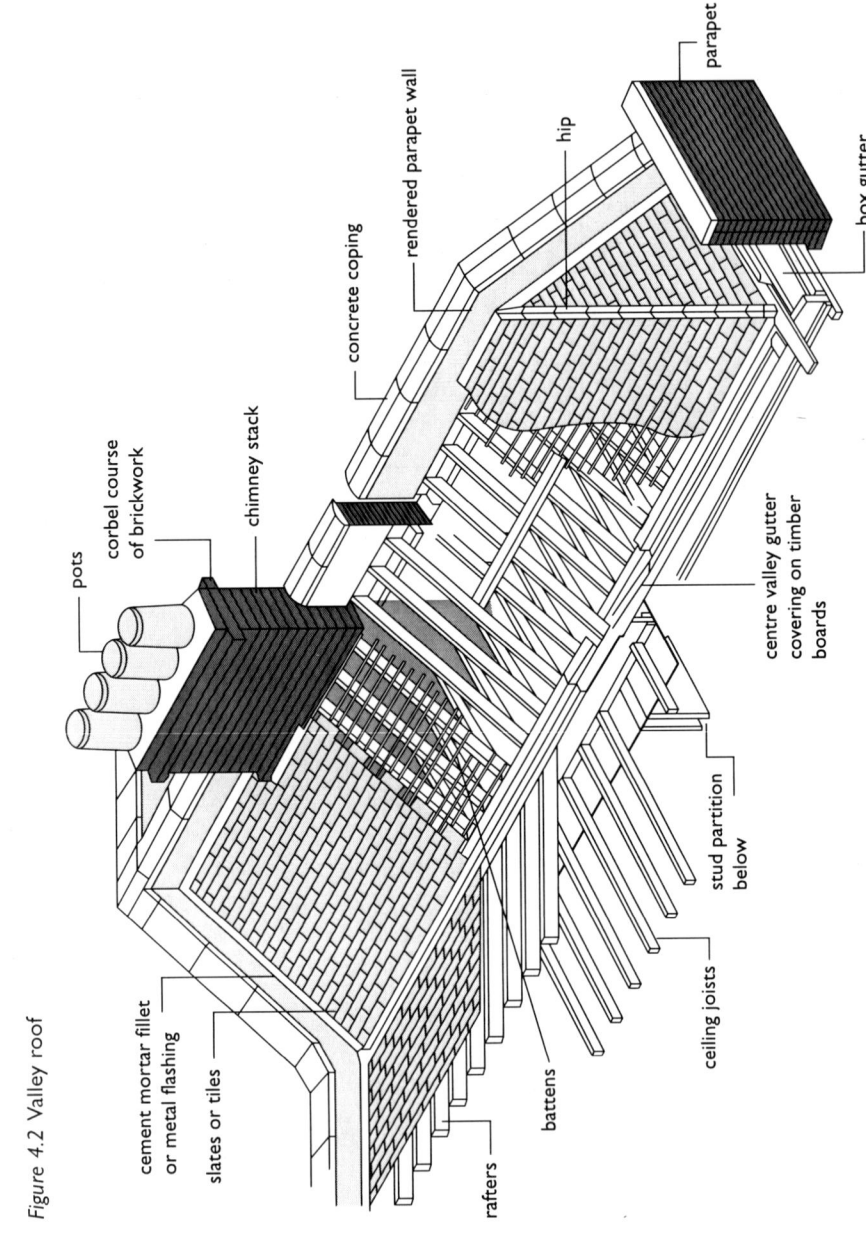

Figure 4.2 Valley roof

Figure 5.1 Metal flat roof

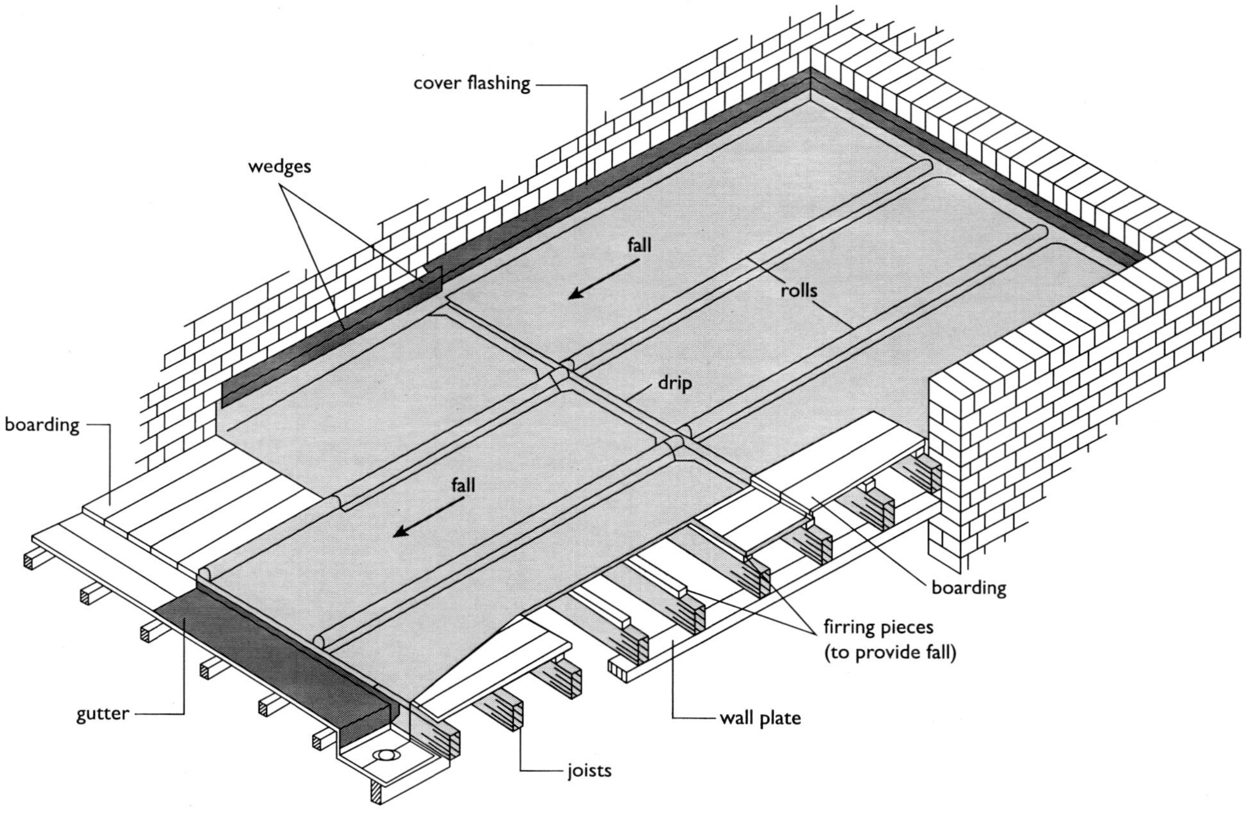

Appendix 1

Figure 5.2 (a) Asphalt and felt coverings on timber roof

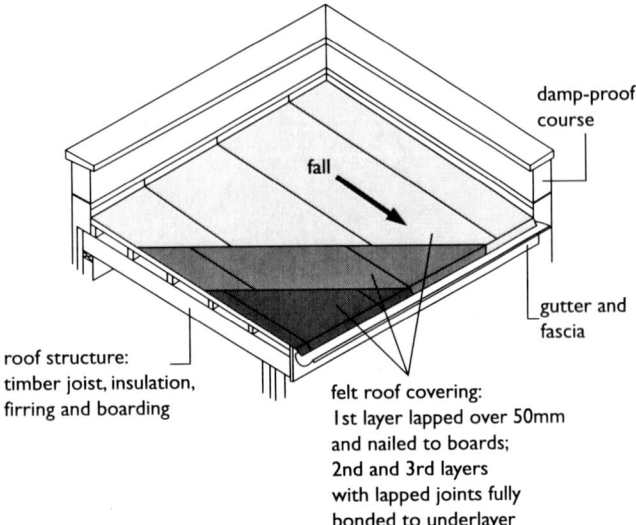

Figure 5.2 (b) Asphalt and felt covering on concrete roof

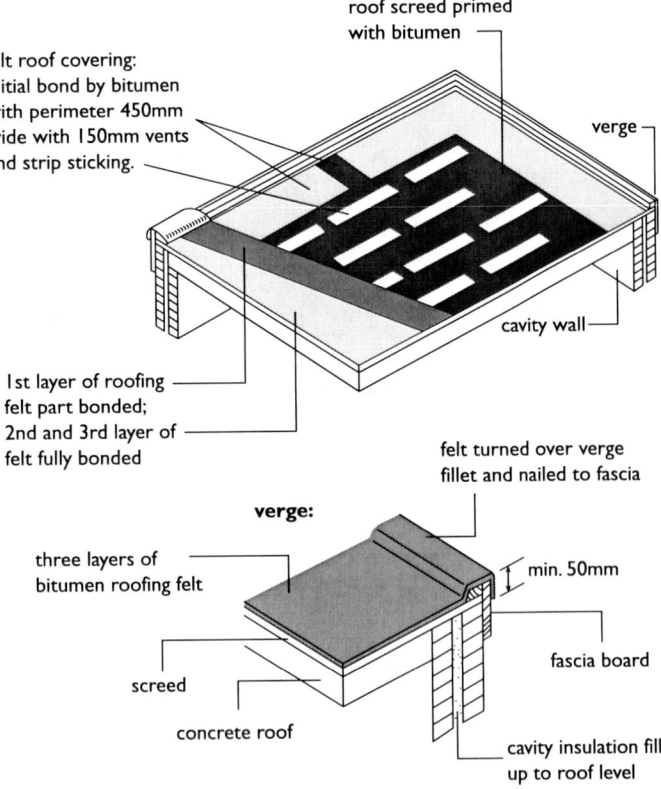

Appendix 1   117

Figure 6.1 Timber upper floor

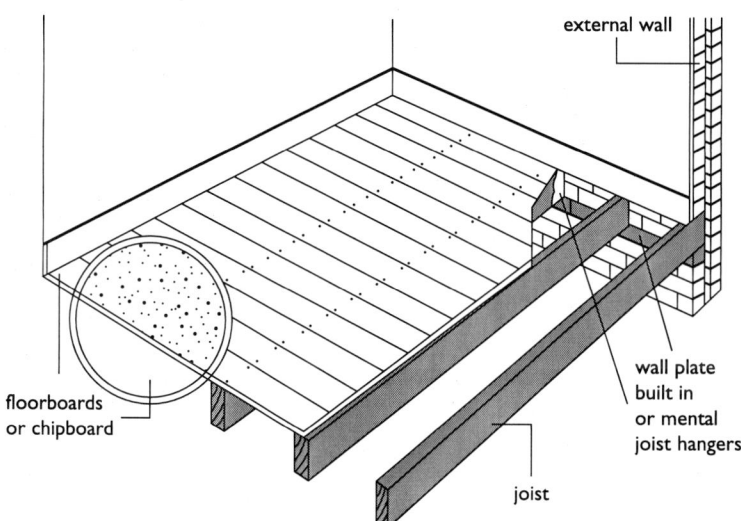

Figure 6.2 Timber ground floor

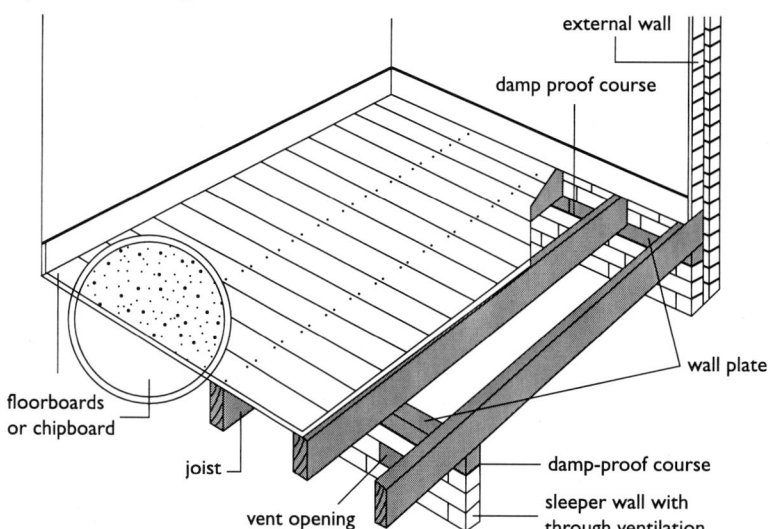

118 Appendix I

Figure 6.3 Solid ground floor

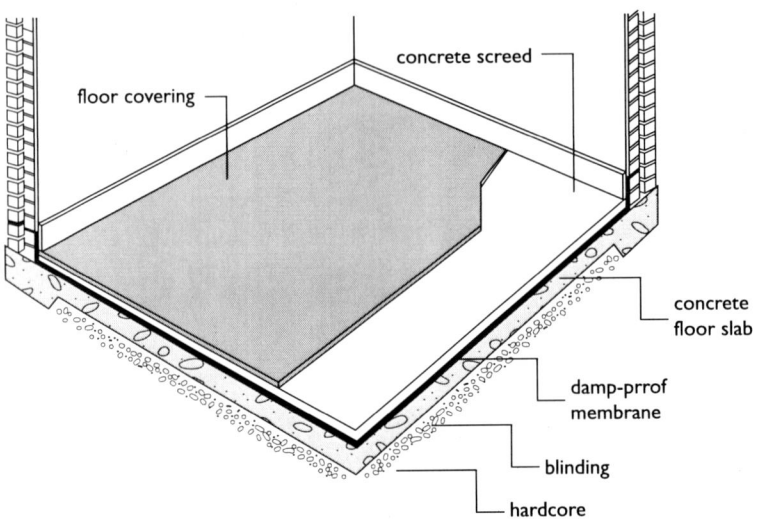

Figure 7.1 Sliding sashes
These slide up and down, but more modern types particularly with plastic frames may slide from side to side. Traditional timber sash windows are operated by weights attached to sash cords. Moderns designs have spring-action spiral sash balances. Some tilt inward for easy cleaning.

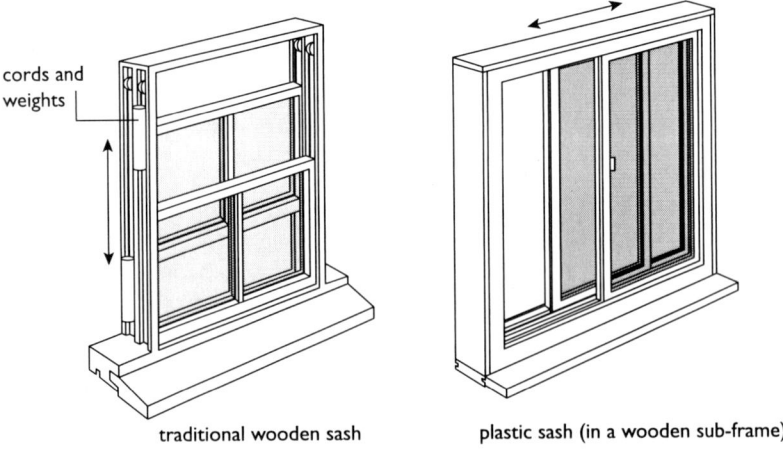

Appendix 1   119

Figure 7.2 Casement windows
Casements open on hinges. They usually open outward. There may also be one or more fixed panes. Side-hinged casements are usually large, while top-hinged casements are small.

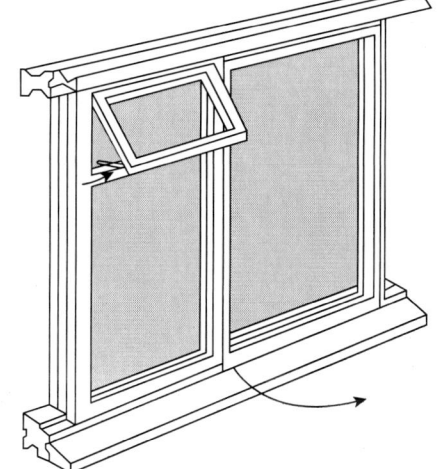

Figure 7.3 Bow windows
Usually multiple casements made to form a shallow curve.

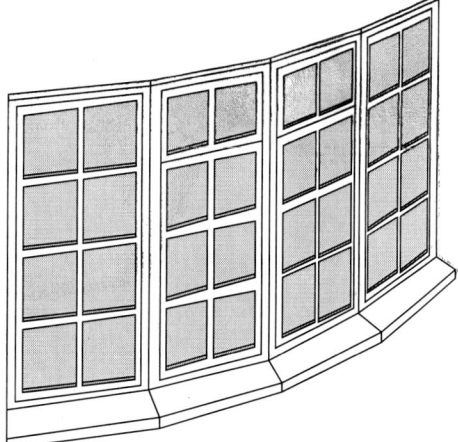

Figure 7.4 Bay windows
Most bay windows are made with stone or brick piers projecting beyond the line of the wall, and fitted with sliding sashes. But a bay window can also be made with a series of casement or sash windows joined together.

structural posts at corners

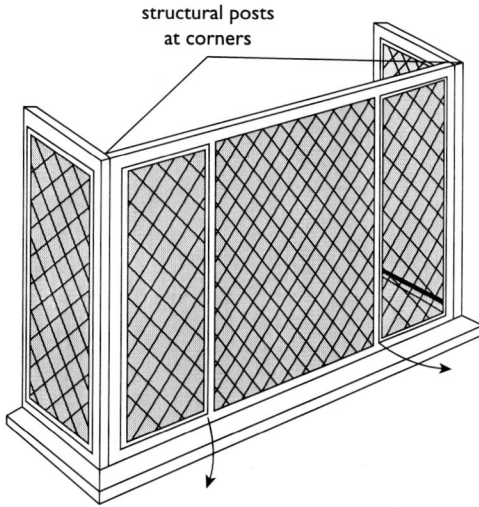

120 Appendix I

Figure 7.5 Louvre windows
Horizontal slats of glass attached
to the frame at each side can be
adjusted to control ventilation.

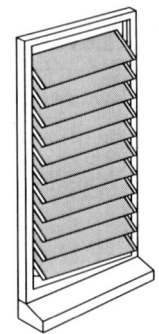

Figure 7.6 Pivot windows

Projected windows.
A sliding action moves the
bottom of the window
outward while the top
rail slides down in channels
in each side of the frame.

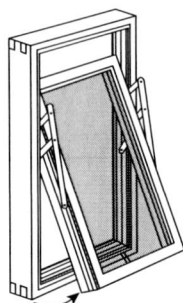

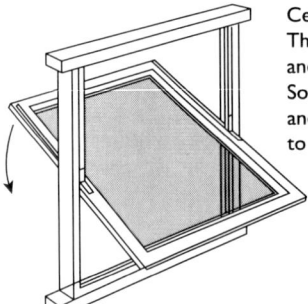

Centre pivot.
The window pivots at its mid-point,
and may also have a fixed pane.
Some types are reversible for cleaning
and maintenance. Safety catches are fitted
to prevent accidental rotation.

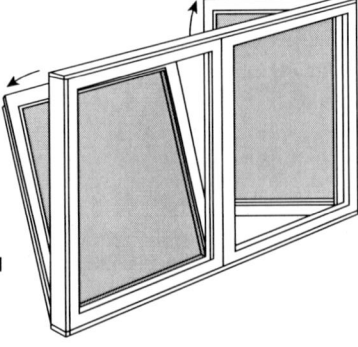

Tilt-and-turn.
These open inward at the top, for
ventilation without a security risk, and
also open from one side for cleaning
and maximum ventilation.

Appendix II

Diagnosing building defects: 'following the trail'

'Following the trail' is a term taken from surveying practice. It is often referred to in litigation and concerns surveyors' failure to observe or adequately interpret 'trigger' defects.

This Appendix does not give the reader all the possible options which could cause or result from each trigger defect as these are available from other sources, including computer information programs. It does give an example of the process and sets out the first few questions to reduce these options to a manageable number.

Appendix II

Diagnosing building defects

Trigger defect	Initial options of cause/result	Question/answer	Options remaining
Window does not operate fully	Jammed shut	Has it been painted?	Yes – stuck with paint No – swollen/rotten or rusted
	Rotten/rusted	If a timber window, is it rotten, or if a metal window, rusted?	Yes – repair window No – other defects

Diagnosing building defects

Trigger defect	Initial options of cause/result	Question/answer	Options remaining
Rising damp	Dampness penetrating from high ground	Is the floor below ground level?	Yes – bridging or lateral penetration No – contaminated plaster
Damp patch on a wall	Water penetration	Is it an outside wall?	Yes – water penetration No – leakage
		Is there a leaking rainwater pipe or gutter outside?	Yes – likely cause No – other discrete cause
	Condensation in a chimney	If it is on a chimney breast, is the flue capped and swept?	Yes – condensation or other source No – water from flue and/or dampened debris on chimney throat
	Condensation in the room	Is there mould growth?	Yes – clean water source, e.g. internal plumbing or condensation No – contaminated water source, e.g. groundwater

Appendix II   123

Diagnosing building defects

Trigger defect	Initial options of cause/result	Question/answer	Options remaining
Water on window frame internally	Condensation	Running water on glass?	Yes – condensation No – water penetration
	Water penetration	Holes around window?	Yes – water penetration No – condensation
Damp patch on ceiling	Water penetration	Roof over?	No – plumbing leak Yes – roof leak
	Plumbing leak	Rooms or loft above?	Yes – plumbing leak No – other water source
	Condensation	Loft above?	Yes – possible condensation from within roof space No – other water source
Damp solid ground floor	Rising damp	Distinct pattern, uniformly damp, high moisture meter readings more noticeable at perimeter	No – leakage or condensation Yes – rising damp
	Condensation	General dampness, low moisture meter readings, dampest under impervious floor coverings	No – leakage or rising damp Yes – condensation

Diagnosing building defects

Trigger defect	Initial options of cause/result	Question/answer	Options remaining
Damp solid ground floor	Leakage	Distinct pattern, uniformly damp, high moisture meter readings, area affected in straight line/following pipe runs	No – condensation or rising damp Yes – leakage
Dampness to walls at lowest floor level	Rising damp	Distinct pattern, uniformly damp, high moisture meter readings, clear cut-off to almost nil moisture meter reading	No – condensation Yes – rising damp
Rising damp	Dampness rising from the ground in the wall	Is there a damp-proof course in the wall?	Yes – bridging or lateral penetration or contaminated plaster No – rising damp
	Dampness rising from the ground in the floor	Is there a damp-proof membrane in the floor?	Yes – bridging or lateral penetration No – rising damp
	Dampness rising from the ground in the wall and floor	Is there a link between the damp-proof course and the damp-proof membrane?	Yes – contaminated plaster No – bridging or lateral penetration

Index

Note: Page numbers followed by 'f' refer to figures and followed by 't' refer to tables.

A
advice, external 79
animals 38
ants 35
appointments, arranging 54–6
asbestos 36–7, 36t; removal contractors 37
asphalt 24–5, 115f
audit trails 91–2

B
background information in preparation for inspection 49–51; address and identification code 49; building type 49–50; common parts 50–1; occupants 50; size and location 49; tenancy agreement 50; type of property 49
basements, damp-proofing 14, 28
blockwork walls 9, 105f, 107f
board flooring 13, 116f
brick 20; cavity walls 8–9, 105f; footings 10, 108f; solid walls 8, 105f
buildings: diagnosing defects 120–2; diagrams 105–19

C
cameras 52
casement windows 15, 118f
casual call systems 53–4
cavity walls 8–9, 105f; damp-proofing 31, 105f; insulation 21, 31
Chartered Institute of Housing Standards 84, 85t
City Challenge 86, 87
cockroaches 35
codes of practice 83
common parts, inspection of 50–1
compensation, tenants' right to 84
complaint records on tenant dissatisfaction with repairs 92–3
computer systems 48, 92–3
concrete 22; foundations, reinforced 10, 110f; frames, reinforced 15, 41; no-fines 16
condensation 23, 31–2; interstitial 32
condensation-caused dampness: energy improvement works to eradicate 94–5; example of 'following trail' to 46; landlord's repairing obligations 78, 80–1
contractors, performance monitoring of 91–2
Control of Asbestos Regulations 2012 36
court proceedings 96–101; choice of expert 99; compliance with orders 101; expert evidence and advice 97–8; fees for expert witnesses 99–100, 103; housing manager's evidence 100–1; instructing an expert 96–7
customers 71

D
damp: diagnosing defects as a result of 121–2; from ground 27–30; lateral penetration 28; moisture generation by users 38; moisture meters 52; rising 27–8; salts and residual 29–30 *see also* condensation-caused dampness

damp-proofing: basements 14, 28; bridging 29; failure of courses 27–8; ground floors 14, 29, 116f, 117f; remedial 28–9; walls 14, 31, 105f
deathwatch beetles 34–5
Decent Homes Backlog Programme 3
Decent Homes Programme 3
diagrams, building 105–19
diary systems 54–5
do-it-yourself 67
DR v Southwark LBC 78
drones 52
drought 39
dry rot 33

E

enemies of healthy buildings: animals 38; fungi 32–4; hazardous materials 36–7; insects 34–5; metals 35–6; non-traditional buildings 40–2; plants, trees and bushes 39; refuse 38; sulphates 32; temperature and climate 39–40; underground threats 40; users 38; water 26–32
energy efficiency 2, 94, 95; standards 3
Environmental Protection Act 1990 81; prosecutions under 96
equipment for inspection 51–2
European Standard of Quality Management Systems (ISO 9001) 1, 102
expert: choice of 99; evidence and advice 97–8; instructing an 96–7; witness fees 99–100, 103
external advice 79

F

fees for expert witnesses 99–100, 103
felt 24–5, 115f
fire walls 11, 24, 112f, 113f
flat roofs 12, 24–5, 114–15f
fleas 35, 38
flooding 26
floors 13, 116–17f; above ground level 13, 116f; board 13, 116f; damp-proofing ground 14, 29, 116f, 117f; sheet 13, 116f; solid ground 13, 117f; timber 13, 116f
follow-up action 91–5; on ineffective or outstanding repairs 95; longer-term 94–5; management systems 91–2; monitoring performance 91–2; monitoring short-term repairs 95; redecoration 93; tenant satisfaction surveys 92–3; variations of the works 93–4
'following the trail' 45–7; and diagnosing building defects 120–2; example 46
foundations 9–10, 108–11f; affected by water 26–7; brick footings 10, 108f; piling 10, 111f; raft and other reinforced concrete foundations 10, 110f; strip foundations 10, 109f; timber plates 9, 108f
frost 39–40
fungi 32–4; dry rot 33; moulds 34; wet rot 33–4

G

glass 23–4
glass fibre 37
Good Housing Leads to Good Health 89
government funding decisions 87

H

hazardous materials 36–7; asbestos 36–7, 36t; glass fibre 37; radon gas 37; urea-formaldehyde foam 37
health and safety 55
high-rise flats 15; defects in 41–2
house mites 35
Housing Act 2004 3, 81
Housing Disrepair Legal Obligations 79
Housing Fitness Standard 81–2
Housing Health and Safety Rating System (England) Regulations 2005 (HHSRS) 81–2
Housing Health Cost Calculator (HHCC) 83
Housing Management Standards Manual 84
Housing: Proportionate Dispute Resolution: An Issues Paper 45
Housing (Scotland) Act 2010 83
housing stock 7–17; non traditional 15–16; traditional 7–15

I

Ikarian Reefer (1993) 97
information: recommended sources of 89; recording 48–9 *see also* background information in preparation for inspection

126 Index

insects 34–5
inspections, disrepair 53–67; arranging appointments 54–6; casual call systems 53–4; conducting 56–7; diary systems 54–5; do-it-yourself 67; example of inspection 64–5; health and safety 55; identifying defects 58; introductions 56; methodical procedure 57–8; notes 58–66, 59–63t; preliminaries 56–7; timescales for repairs 67; using senses 58
insulation 31; cavity wall 21, 31; glass fibre 37
ISO 9001 (European Standard of Quality Management Systems) 1, 102

K
Kenning v *Eve Construction* (1995) 98

L
Landlord and Tenant Act 1927 88
Landlord and Tenant Act 1985 77, 80, 81
landlord's liability 77
large panel systems 16; defects in 41
London & Leeds Estates v *Paribas* (1993) 98

M
management systems 91–2
materials of construction 18–25; asphalt, felt and other bituminous materials 24–5; brick 20; concrete 22; glass 23–4; insulation 21; medium density fibreboard (MDF) 25; metal 24; mortar and pointing 20–1; plaster 22; render 21; rock 21; slates 23; stone 22; tiles 23; timber 18–19
medium density fibreboard (MDF) 25
metals 24; corrosion 35–6; reacting together adversely 36
moisture generation 38
moisture meters 52
mortar and pointing 20–1
moulds 34

N
National Home Energy Rating scheme 94

no-fines concrete 16
non traditional housing stock 15–16; enemies of, and defects in 40–2; high-rise 15, 41–2; large panel systems 16, 41; modern timber frames 16, 42; no-fines concrete 16; prefabricated 41; steel and reinforced concrete frames 15, 41
notes, inspection 58–66, 59–63t

O
occupied dwellings 47
orders, compliance with 101

P
performance indicators for repairs 83
performance monitoring of individual contractors 91–2
pets 38
piling 10, 111f
pipework: corrosion 30, 36; leaking 30, 40
pitched roofs 10–12, 112–13f
planned maintenance 84, 86
plants, trees and bushes 39
plaster 22; repair obligations 77–8
pointing and mortar 20–1
prefabricated buildings, defects in 41
preparing for disrepair inspections 45–52; background information 49–51; cameras 52; deciding on necessity of an inspection 47–8; equipment 51–2; 'following the trail' 45–7; housing manager's role 45; moisture meters 52; occupied dwellings 47; recording information 48–9; starting points 47; third parties 47; use of drones 52; voids 47
priorities, repair: assessing 80–9; bringing forward a repair for cost-effectiveness 89; codes of practice 83; deferring repairs for more cost-effective renewal 88–9; external advice 79; general factors 80–1; planned maintenance 84, 86; specification for repairs 80; statutory obligations 81–3; Tenants' Guarantee 83; Tenants' Right to Repair 84; value for money 88–9; wholesale redevelopment or rehabilitation 86–8
procedures 102

R

radon gas 37
record notes 76, 76t
recording information 48–9
redecoration 93
redevelopment or rehabilitation schemes, wholesale 86–8; assessing validity of landlords' intentions for 88; funding 86, 87; similarities with commercial properties 88
refuse 38
reinforced concrete and steel frames 15, 41
render 21
replacement windows 15
reporting: core section 73, 73–5t; to customers 71; of findings 71–2; format 72–8; landlord's liability 77; record note 76, 76t; reference to further action 78; sample schedule of defects and repairs 73–5t; supplementary sections 76; tenant's liability 77
resources, efficient use of 102–3
rising damp 27–8
rock 21
rodents 38
Ronan Point 41
roofs 10–12; chimneys 11, 112f, 113f; fire walls 11, 112f, 113f; flat 12, 24–5, 114–15f; front to rear pitched 11, 113f; hips, valleys and gables 11, 112f, 113f; parapets 11, 112f, 113f; pitched 10–12, 112–13f; slates 11–12, 23, 112f, 113f; tiles 12, 23, 112f, 113f; valley 11, 113f
rot: dry 33; wet 33–4

S

safety glass 23
salts 29
schedule of defects and repairs, sample 73–5t
settlement 39, 40
sheet flooring 13, 116f
short-term repairs, monitoring 95
slates 11–12, 23, 112f, 113f
sliding sash windows 14, 117f
snow 40
social housing standards, current criteria for 3
solid brickwork walls 8, 105f; insulation 21
specification for repairs 80
statutory obligations 81–3
steel and reinforced concrete frames 15, 41
stone 22
strip foundations 10, 109f
stud partitions 9, 106f
subsidence 40
sulphates 32
sun 40
surveys, tenant satisfaction 92–3
sustainable construction 2
Sutherland Tables 95

T

tanking 14, 28
temperature and climate 39–40
tenancy agreements 50, 77
tenant satisfaction surveys 92–3
Tenants' Guarantee 83
tenant's liability 77
Tenants' Right to Repair 84
third parties 47
tiles 12, 23, 112f, 113f
timber 18–19; -boring insects 34–5; floors 13, 116f; frames, modern 16, 42; plates 9, 108f
timescales for repairs 67
traditional housing stock 7–15; damp-proofing 14, 105f, 116f, 117f; external walls 8–9, 105f; floors 13, 116–17f; foundations 9–10, 108–11f; internal walls 9, 106–7f; roofs 10–12, 112–13f; windows 14–15, 117–19f
trees and bushes 39

U

underground threats 40
urea-formaldehyde foam 21, 37
users of buildings 38
Using the Housing Act 2004 89

V

value for money 88–9
voids 47

W

walls: blockwork 9, 105f, 107f; cavity brickwork 8–9, 105f; damp-proofing 14, 31, 105f; drying out of residual dampness 29–30; external 8–9, 105f; internal 9, 106–7f; solid brickwork 8, 105f water 26–32; from above 31; above ground level 30; from construction 32; dampness from ground 27–30; from inside 31–2, 38; underground streams 40; ways an enemy of buildings 26–7

wet rot 33–4

windows 14–15, 117–19f; casement 15, 118f; diagnosing defects 121; replacement 15; sliding sash 14, 117f

witness statements 100–1

wood boring insects 34–5

Wood Moisture Equivalent (WME) 52

works, variations of 93–4